"We experience a constant barr[illegible]
some 'new scientific study' has p[illegible] be-
havior and experience has been isolated to a particular part of the brain. The implication often is that brains exist but 'we' (as conscious, whole persons that are somehow more than electrochemical impulses) do not. Mark Cosgrove's book unpacks what we know of the vast complexity of the human brain while celebrating the reality of the person who can stand back and decide to explore such gloriously complex questions. He explains thoughtfully and effectively how neuroscientific advances in understanding the complexity of the biological structures and processes of the human brain, when properly understood, inform and expand our understanding of personhood without destroying it. I highly recommend this engaging and wide-ranging exploration of the complexity of the brain and mind."

—Dr. Stanton L. Jones, author of *Psychology: A Student's Guide*, coauthor of the *God's Design for Sex* book series for families, and coeditor of *Modern Psychotherapies: A Comprehensive Christian Appraisal*

"With prayerfully discerned wisdom, Mark Cosgrove offers a compelling yet accessible companion to all wanting to wade into the depths of neuroscience. Experts and novices alike will profit from no longer seeing the mind in functional isolation but as a critical partner in what it means to be fully human and thus fully alive."

—Todd Ream

"*The Brain, the Mind, and the Person Within* is a remarkably gentle and humble introduction to consciousness studies. Cosgrove never fails to keep in focus the real object of study in this field—persons. Somehow he manages to make even neuropsychology and philosophy of mind light and engaging. This is no mean feat, given the technicality of those disciplines. I recommend this book both to novices in consciousness studies and those well-versed in the field who would like a refreshingly warm and human discussion of the subject."

—James S. Spiegel,
author of *The Benefits of Providence* and *The Making of an Atheist*

"Finding the mind inside the brain…will we ever reach that frontier?"

—Jeff Van Valer, M.D., neurologist, sleep physician, and author of *The Devil's Tricycle: Migraine Headache, Caffeine Abuse, and Insomnia*

MARK COSGROVE

THE BRAIN, THE MIND, AND THE PERSON WITHIN

The Enduring Mystery of the Soul

The Brain, the Mind, and the Person Within: The Enduring Mystery of the Soul

Published by Kregel Publications, a division of Kregel Inc., 2450 Oak Industrial Dr. NE, Grand Rapids, MI 49505-6020.

ISBN 978-0-8254-4526-2

Printed in the United States of America

18 19 20 21 22 / 5 4 3 2 1

To my beautiful wife, Jo Ann, always at my side;
and my three sons and their wives,
Walker and Kirsten; Robert and Julia; Preston and Jennifer;
and our growing family of grandchildren.
You are all the best books I will ever write into,
and the best books I will ever read from.

Contents

Preface

Albert Einstein is reported to have said that if you cannot explain something simply, then you do not understand it. I want this book to be written for people with all sorts of educational backgrounds, not necessarily in any formal science, but with an interest in human nature and the findings in neuroscience. I want the reader to understand the issues coming rapidly in the neuroscientific arena, to see the value of faith assumptions alongside current assumptions, and research in the study of the human brain.

I want the book to be clear and accurate about science without straying into oversimplification. My overall goal remains to help the reader understand the issue or the science involved with the human brain without being too technical in my writing. For the reader's interest, I have added a reference list of books on neuroscience at the end of this book. Most of the books are very readable and deal with the topics I have included in this book. The authors are interesting thinkers and researchers. I have also included a smaller annotated reference list on books, some with a decided Christian perspective that many of us share.

A number of different viewpoints are trying to get to the heart of the same mystery, the mystery of ourselves, created with God's spirit, embodied in the flesh, and who will survive the death of our brains. We are one being, spirit and matter fused together, and we become in the Hebrew language *nephesh* or living being, until death, and then we await the resurrection of our bodies and brains.

CHAPTER 1

THE OLDEST BRAIN

As I sit at my desk I am looking at a picture from a science report on the oldest brain ever discovered. A team of scientists from the York Archaeological Trust discovered in 2008 a skull with a brain inside. The skull, which was found in an Iron Age site in the UK, has been dated to the sixth century BC. That makes it 2600 years old! Amazingly, the brain is still intact inside the skull! Its soft spongy parts did not decay because the person was decapitated and buried face down in the mud. The brain appears like yellow spongy material. I find myself asking, *who was in that skull*. Was he afraid as he was killed and buried immediately in that spot? And I wonder how old he was, and why he had to die. He was a person of worth and value as we all are. I am thinking a bit like Hamlet as he looks at Yorick's skull and ponders about a familiar life that once was in that court jester's skull and now is not. What is it about human beings with mere three-pound brains who must engage in such heavy thoughts?

A TITLE CAN SAY A LOT

The Three-Pound Universe

—*Hooper & Teresi*

THE HUMAN BRAIN: AN INTRODUCTION TO A MYSTERY

"If the human brain were so simple that we could understand it, then we would be so simple that we couldn't."
—Emerson M. Pugh

Atlas, the mythological Titan, is often seen in pictures, bending under the weight of the earth as he holds up the globe on his muscular shoulders. Zeus condemned Atlas to hold up the heavens for daring to make war against the Olympian gods. Lifting the whole planet is a load that none but an Atlas or a Hercules could possibly bear. However, we mere humans, small or large, weak or strong, with just our three-pound brains, manage to hold up the entire universe in our minds, on which to see and reflect. Our three-pound brain with its pinkish-gray surface of cells, as fragile as jelly, does not just move our legs, chew our food, and manage our reproduction. This mass of cells inside our skulls amazingly allows us to hold up and reflect on both the largest and the smallest parts of the universe and everything in between.

With our human brains of approximately 86 billion wet, neural cells, according to a recent count,[1] and one trillion smaller glial cells, you and I can almost see the ends of the universe, thirteen-and-a-half-billion-light years away through the lens of the space-traveling Hubble Telescope. Only a few thousand light years from earth, we can see in color the beautiful Hubble pictures of the Pillars of Creation in the act of giving birth to untold numbers of stars. That Hubble picture of those majestic pillars of stellar clouds was first shot in 1995 and then reimaged in 2015 with newer infrared cameras. The spectacular color image, looking like a hand with fingers pointing upward, makes us feel the presence of God since the starry heavens are spoken of in Psalm 19:1 as the "handiwork" of God.

With our same three-pound brains, and a change of camera and focus, we can also view the infinitely smaller stuff of our universe with the Hadron Super Collider on the French-Swiss border. The super collider name is accurate since this machine is a super atomic blaster. The LHC, as it is known, is the most complicated machine ever built. This human hammer of Thor operates three hundred feet underground and has a 17 mile, circular, atom-smashing tube crashing subatomic particles together at lightning speeds (99.99999 % the speed of light!) and with thunderous energy, exposing even smaller, subatomic particles yet undiscovered by human beings. We can hold up for inspection this enormous picture with just three pounds of our leaking and sparking brain matter using less energy than a refrigerator light bulb. With such a close-up view of the tiniest parts of the universe, we say we can witness creation. With this focus on a foundational particle, scientists named the newly discovered Higgs Boson in the Hadron particle accelerator the "God" particle. Such heavy thoughts from just three pounds of brain matter! Interestingly, one of the experiments with the Large Hadron Collider is called the ATLAS experiment, looking for what holds up the supersymmetry fabric of our universe. Our interests in such projects makes us wonder what universal drive is up there inside our brains that is of such importance beyond the urge to eat and breed.

The human brain is not much to behold when you are holding it in your hands, or viewing it with its dead-white color, floating in a jar of formaldehyde. But when that brain is in the skull of a human being, look out! Alive, the brain becomes what we call a person, capable of deep emotions, amazing scientific discoveries, tender poetry, cruel behaviors, and love of other human beings and even God. That brain, now a person, does not just build cathedrals and compose love poetry, but it is consciously aware of what life feels like, of falling in love, of seeing red, and hearing middle C played on a piano. That brain, no, a person, is capable of feeling meaning, awe, joy, hate, beauty, truth, heroism, honor, guilt, and humor, and that person fills her paintings and novels with the same rich emotions.

As I describe some basic features of the brain, let me suggest that nothing I say about personhood should take away from the wonder of the human brain, because that brain is the embodied person, who is you. I am describing nature, but I am also describing you. And, the pieces of brain I will describe are unified, and are not just machine brain part names to memorize for multiple choice exams. We should train ourselves to see the brain and its functions as much as a dance as a complicated biology. Then, with a poetic eye, we can begin to see meaning and beauty in brain and our embodied selves. Dancing in a biology lab might make some silly sense, I suspect.

The self-consciousness mystery in our skulls, which is us, is fused to the activities of those 86 billion neural cells in the brain, each of which makes five to ten thousand connections with other neural cells. The trillion even tinier glial cells, some called stellate or star cells, surround the neural cells and make contact with anything in range, allowing more interconnections between brain cells than there are stars in the entire universe, far more than Hubble could ever view! The mystery of our conscious, thinking, feeling minds deepens with every brain scan we perform to find out more about this three-pound enigma. The facts we gain with the activity scans like the familiar PET and fMRI scans make us feel like we are just standing in ankle-deep water in the ocean and seeing how much more there is to explore. Indeed, with every step into the research, the waters get deeper with more mysteries to unravel. There is so much to understand in this mystery of three pounds. We have learned to listen in on and initiate the activities of individual cells with our probing electrodes and micro pipettes. We study the rhythms of the brain with our EEG recordings as a person sleeps and thinks. We can lay down in an fMRI machine and watch where our brains are more active when we pray and when we laugh.

Step Back and See

In looking at a piece of art, one has to often step back from the framed picture in the museum in order to take in the whole of the picture and see what is being portrayed and felt on the canvas. Likewise, if you look at a map and want to see where you are, you might have to draw your gaze back a bit and look at a larger area on the map from where you think you are, and spot a few familiar landmarks or streets. The same seems true in our study of the brain and what has been called by most neuroscientists, "the most complex physical structure in the universe." We often need to pull back from the tiny bits of scientific data and see how our data fit in with the whole of other data and theories, especially since we human beings are the data being examined. That has not been the approach of neuroscience, as repeatedly the field often ignores larger views of human beings, and other ways of knowing than its empirical studies. The best approach to knowing what a painting is about, or where you are in a city, is not to look at every little color dot, or to look at every street sign, but to view the larger picture within which the smaller pieces of information will fit and be understood. You have to do both, see the big picture and look at the details. Looking at the details is referred to as bottom-up thinking and research. And, seeing the bigger picture is seeing from a top-down, higher view.

Radical empiricism is a way of knowing practiced by some scientists in which it is said that I do not need anything but knowledge by sensory experience and the methods that flow from that. Empiricism is a hugely successful way of knowing, and I believe I can know through my senses. More radical empiricists, like B. F. Skinner, of behaviorism fame in the middle 1900s, argued in favor of sensory empiricism studying behavior not the mind. No subjective reports, please, just objective information. A radical empiricist says to us, do not tell me about your feelings; I will gather objective data and put together the whole picture of you, and tell you what to believe about yourself.

Opposed to such radical empiricism, will be the approach that says keep in view the concept of personhood that we experience in ourselves and in our interactions with others while we study the details of the brain empirically. Do not assume that such top-down thinking means that religion or angry popes will be telling the scientist what is true in the details of the science. Such top-down viewpoints may indeed speak too forcefully at times, but the brain sciences should be open to stepping back and seeing the larger world of human experience and thoughts as helpful guides to research directions and to the interpretations and applications of research findings.

In the interest of stepping back, in this book I will use examples of persons and their inner lives, and seek to explain personhood and not just explain it away. Such examples help us with the discovery of self and often help in the understanding of the neuroscience data we are unveiling. Many important discoveries about human nature are coming to us because of empirical studies in neuroscience research. Other facets of human nature can be revealed in the depths of poetry, art, music, and religion, and those forms of knowing are every bit as useful for understanding the complexities of human nature. Opening up ways of knowing in neuroscience seems like the wise thing to do because the subject matter, which is us, sits on the edge of matter and mind. We are hybrid creatures, embodied spiritual beings, and the one piece of creation in this natural world that is able to look up and reflect on this universe and our place in it. To call us freaks of nature ignores the personal examples of great genius and humanness that I will bring up in every chapter. This book is about neuroscience, and so I will be primarily discussing the brain and the research on it. However, I never want us to forget that there is more that is out there in the world of our common experience, which is related to our brain's activities, and connected to the world of matter and ocean and star and beyond. We will back up in every chapter with a look at some fantastic individuals, who have minds and personalities so much beyond the explanations of brain function that fill our textbooks.

It would be less than humble for me to be too specific about how I think the brain and/or mind of the person is unified and functions. Almost every neuroscientist seems to agree that the human brain is the most complicated thing we will ever find in the universe, and if we add a spiritual dimension to such brains, then our personhood is even more complex. I will discuss some theories of mind and brain in this book, and no apologies are needed to say that our theories may not be at all adequate to describe ourselves as persons. What I do want is to never lose sight of ourselves as persons, spirit and matter tightly knit together, whatever that means, and to move our scientific research in that direction.

The Pooh Bear Problem

As successful and skillful as we have been with our examination of the human brain, there is an obvious difficulty in studying the human brain and the mental life of the person. We are trying to understand ourselves using just ourselves and our own minds. I saw a Winnie-the-Pooh Bear cartoon once that showed Pooh Bear scratching his head as he looked at a toy pooh bear stuffed animal in his hand. The caption read "Will Pooh Bear ever be able to understand pooh bear?" In other words, how can Pooh Bear ever understand himself using just his pooh bear brain? The Pooh Bear caption was asking the same question as a famous 16th century woodcut from the first edition of *De humani corporis fabrica* (on the workings of the human body) by Andreas Vesalius, which shows a human skeleton studying a human skull on a table. This book was first out in June of 1543, just days after Nicholas Copernicus had published his *On the Revolutions of the Heavenly Spheres*. The skeleton in the woodcut is leaning on its bony elbow as it stands at a table, peering closely at the skull. That woodcut and Pooh Bear make us ask ourselves how can we humans ever hope to understand our own brains, our purpose, and meaning in this immense universe, using only our own brains with which to investigate such things?

The answer often given is that we, like Pooh Bear, will never be able to understand anything beyond our material brains and our immediate environment. We are told not to worry about this limitation because that is all there is to you, a material brain of complicated mechanics. Reductionism, the philosophy of reducing all things to mere matter, says to stick to objective observations and do not make pronouncements beyond the brain sitting on the table in front of you. What is interesting, though, is that we, biological humans beings, do understand so much about ourselves and the universe in which we find ourselves, and more is opening up to us all the time. We appear to be so much more than the mere matter of our brains, and our discoveries with Hubble and LHC and our understanding of much in this vast universe are evidences of that. The fact that we have motives for searching out the meaning of the universe and its underlying physical framework seems to argue for more to us than mere brain cells. Why else do we long for more than bananas and grass in a universe that is supposedly mere matter?

We do understand so much of the brain, with our little Pooh-like, three-pound brains. We understand much of life, and of meaning, perhaps because the brain we examine in our heads is no mere brain, as we will see later. Those who believe in God say human beings are a union of brain matter and spirit from God. Call us a biological accident and a freak if you want to, but you are pushing a boulder up a steep hill. As good as a goldfish's color vision is, we do not expect it to understand what is on a color television set 12 inches from his bowl on Super Bowl Sunday. That is because seeing is more than generator potentials in rods and cones. What can the goldfish make of a football bowl game played a thousand miles distant from its little bowl of water? The game's image is carried on invisible waves, showing men who have hopes, dreams, wives, and families. If goldfish could think, and had their own schools of learning, what theories could they possibly develop as explanations for what they saw outside of their water worlds? Would they even want to? We, on the other hand, do not have that same, forever-inaccurate view of reality. The truth is, we understand so much more of what is out there calling to our deeper desires to know.

Finding an answer to the Pooh Bear problem is greatly improved when one considers that in the Christian worldview it is believed that God communicates truth about humanity to human beings in the Bible. The revealed word of God in the Bible describes a view of the human person that fits the biblical record of matter and spirit fused together to create a unified, living being, a person. We believe that we survive the death of our brains, are resurrected with new bodies and brains, and have eternal value and purpose in God's eternity. We lean on the authority of inspired Scripture, and trust the writings of such great minds as the apostle Paul, the church leaders Augustine and Aquinas, and so many more. It has been obvious for such a long time in our own experiences that something more is going on in our three pounds of brain, and it does not have to necessarily be a body-soul dualism as pictured by Arthur Koestler's oft quoted book title, *The Ghost in the Machine.* A long line of poets and philosophers, priests and pastors, novelists, and scientists continue to voice this view of personhood down to this day. And never has a time needed such a personal viewpoint more, when so much depends on thinking clearly about science and the nature of human persons.

Oliver Sacks—A Person of Interest

We should step back and ponder the brain of Dr. Oliver Sacks, who died in August of 2015 at 82 years of age. He invested his professional time as a neurologist, seeing the personal in his patients, and that viewpoint allowed him to see more than just damaged brains. In his books, he viewed personhood for all of us. Dr. Sacks had a keen scientific mind, but he behaved more like a great, soft teddy bear around persons suffering from neurological problems. He became a gifted communicator, not just to his book audience, but especially to his patients as he showed a compassionate interest in them and their lives. He was poetic in describing people's lives and that allowed us to see more in their conditions. He examined people, not with his empirical hammer, but with a paint brush of compassion, and he searched for their human nature instead of their machine nature. Dr. Sacks' work was a challenge to how neuroscience should be conducted and communicated.

Sacks' vast reading audience began with his book *Awakenings* in 1973 (later a movie in 1990 with Robin Williams as Sacks). The story was Sacks' 'compassionate look at individuals paralyzed by the influenza virus and presumably not conscious of much for decades. Sacks tried L-dopa, a precursor chemical to the brain-transmitter substance Dopamine. The patients gradually came out of their paralytic stupor and began to reclaim their lives in the hospital under Sacks' care. Sacks reveals the tragedy of these people when they later began to revert back to their paralytic state, only this time knowing what was coming, because the medication could only work for a time.

Sacks continued his writings with his best known, *The Man Who Mistook His Wife for a Hat*, a book of neurological patients, whom he described in human terms to help understand their conditions. The patient

Dr. P. was the man who had prosopagnosia, a rare disorder of the brain where the patient cannot see or recognize faces. The title of the book comes from a time when Dr. P. reached for his hat only to grab his wife's head! It is interesting that years later, Dr. Sacks himself suffered from prosopagnosia. Throughout his career, Dr. Sacks described his patients for all of us, such as the brain surgeon with all the spastic movements of Tourette's syndrome. He wrote about a person with only minutes of short term memory, who, with no new memories, lived for years never aging a day in his mind. Sacks described with a biographical friendship the life of the intelligent autistic, Temple Grandin, a professor at Colorado State University and an expert in the treatment of animals. We the readers understood the brain better because we could see it so clearly united with real persons and not just symptoms.

These patients were treated by Oliver Sacks as persons and not as damaged brain machines, which is exactly how we should be studying the brain. The empirical approach by itself to studying human nature can only see what is present in the neuron's sodium and potassium influx and efflux, and synaptic vesicles carrying chemical passports to more neurons. Empiricism, or knowing through our senses, is fine if materialism accurately describes everything in the universe. A radical empiricism of objective data alone may show all when studying the elements of matter, but finding a periodic table of human beings can never describe what we are finding in the experiential life of a person. When studying the brain with personhood in mind, even the empirical data itself can take on new meaning and interpretation. To not view the human brain and experimental data with personhood in view, is a self-imposed poverty of the intellect and greatly limiting in ways that can never help us see the totality of what human beings and their brains are all about.

The Loss of the Personal in the Study of the Brain

It seems true that some of today's science has tried to banish mystery and personhood from human life. Neuroscience in particular, with its subject matter being more complex than any other we can imagine in the universe, seems bent on banishing all language and data that would argue for a larger view of the human brain than mere cause-and-effect activities of matter. Even education seems to have shifted its emphasis from *my what a miracle you are* and *look at the depth of the human mind and emotions in the narratives of life*, to *you are just a more complicated neural version of the C. elegans worm with its 302 neurons for a brain.* A mouse with 75 million neurons in its brain is almost understood, neuroscience claims. Can you, with your nearly 86 billion nerve cells, be very far behind in terms of being explained as a behaving, biological machine? In neuroscience, we are told that science is quickly removing the human brain from the dissecting table and the activity scanning machines and placing your brain on the book shelf with the supposedly understood topics of genetics and life itself. Neuroscience supposedly can now consider itself on par with physics that some say is rapidly explaining the mysteries of the universe with only gravity and subatomic particles yet to be understood. Of course it would help the materialistic cause if the field of physics, the royalty in the sciences, got rid of the squabbles over the necessity of subjectivity in its work with subatomic particles.

Many neuroscientists seem to begin with the assumption that there is no self or mind beyond matter, and nothing beyond mere matter in the universe. Then, using methods that can only find matter, they develop explanations in line with all this. This is circular reasoning and only serves to reinforce a very unscientific start to explaining the human brain and person. Human achievement and complex mental experiences are seen as just made up of smaller pieces of biological explanations, and we will soon see how they all fit into the whole person observed. Whenever some complex behavior seems impossible to answer, some neuroscientists shout, "give us more time and we will have it, we promise." Or, for subjective human experience or feeling itself, seemingly impossible to explain in an objective, material universe, the cry is, "it's just an illusion," from some who think they already know the answer.

Once the assumptions in modern neuroscience have been given the status as proven, scientific facts, then issues dealing with human beings and their problems get tied down to material answers and solutions. The world of the present and the future, with its fractured way of thinking about the human brain and person, is going to have to deal with future ethical questions in mental health or the emerging neuro-technologies with only material guidelines in place. Materialistic views on human nature are having difficulty dealing with ultimate questions such as the value, purpose, and significance of human beings. What serious thinker would pick a method of knowing that could only find what it set out to find? A little humility of knowledge is needed in neuroscience when it studies you, the person, you with your three pound brain that can lift the Hubble Telescope and discover and awe over the widest parameters of the universe.

The Humanistic Renaissance from the 1200s to the 1600s resulted in a confidence in the human mind blossoming in the arts and sciences. That same surge of confidence to know, however, ended up relegating the human mind to the same category as the rock and the chipmunk. Those great humanistic thoughts uplifting human freedom and knowledge were quickly dissolved into a *Clockwork Orange* with no clock maker in the world. Even our great ability to know the world and ourselves has become stuck in an epistemological crisis of us no longer being able to trust our knowledge because, after all, even our words and thoughts are determined. The Renaissance, which began with such lofty thoughts about human potential, eventually put all of us in a box of our construction and shut the lid.

God-of-the-Gaps Problem

One argument against any supernatural view of this personal brain, is called the "God of the gaps" argument. Essentially it goes like this: Whenever science has not explained something about the human mind with biological explanations, Christians say that God did it, or He created some mysterious soul up there in the head of the person. You fill in the gaps with God or miracles, and thereby you can keep clinging to the belief in something still sacred and mystical about the human being. The popular accusation states, "before you say something is out of this world, make sure that it's not in this world."

We Christians need to not fall into the error of believing that somehow God always works on the outside of science, as if science was not also one of God's ways for us to learn about reality. However, realize also that the argument goes both ways when science promises, "just give us fifty more years and we will have the answer." Science, after all, has been so successful in understanding the essence of living things, and the nature of genetics. Can the physical nature of the mind be far behind? This is the promise of science in the future, and for something as important as the study of ourselves, and as difficult as unraveling the brain seems to be, we need to not be so confident. Science is throwing up its own "Science in the Gaps" answer. Science has made amazing progress, yes, but with every rock that we turn over in science, there only seems to be more rocks to lift. Rather than getting finished, science is showing us that the world of nature and the human brain as more and more complicated in every way. We must agree that the human brain is a fantastic mystery, a mystery that science is making progress on every year. Let us also agree that the three-pounds of our brain matter is also strangely us, and we are also that brain up there in the skull. That truth is too difficult to say clearly, but we are clearly living it out every day.

We are all living with the struggle of Atlas, holding up reality with which to contemplate and live in line. That struggle of knowledge is worth the effort, and we all should avoid the too-easy dismissal of anything not our view. The job for us is to be open to both ways here—religion and science. I highly respect the world of neuroscience and its methods and findings, but its conclusions about us as persons are often too much guided by prior assumptions heavy with reductive materialism (reducing everything about the person to nothing but matter). Hubris, the I-know-it-all attitude, often present in science, is never attractive in learning endeavors. Atlas struggled with the globe, and we too should struggle with the universe of facts all around us that we seem built to discover and interpret. Part of that struggle is staying clear of our own biases and pet theories, but not surrendering our world- and life-view as created beings in God's world. We live as persons, seeing ourselves as deeper beings, beyond mere objective, neural/synaptic descriptions, as seen in our moral notions, life after death hopes, and our longings after the meaning of beauty, truth, justice, heroism, and love. This inner self ought to be listened to as we investigate this greatest of mysteries, the mystery of self, from the perspective of the scientist, and the larger picture of a God-created brain—a brain built to lift us higher, to see all that is true and possible in God's realm

An Overview

We will cover many topics in neuroscience in this book, from consciousness and mind/body issues in the first chapters, to free will, God Spots, and robots in later chapters. Neuroscience is unveiling applications from head transplants, mind-to-mind communication, and using thought as remote control over computers, artificial legs, garage doors, and televisions. Developments on the horizon include direct mind links to Twitter and distant music concerts. New brain-chip therapies for depression and other psychological problems are being tested, as well as the planting of or removing of memories and learning from human brains.

We will see the brains of worms with a mere 302 neurons, as well as mouse brains with a hefty 75 million, and of course the human brain with its 86 billion, said to be the most complicated physical structure in the universe. We will see the attempts to map the entire brain as a whole unit. We will see ancient brains, half brains, brains with missing parts, and brains that have been made see-through for better study. We will see how capable brains are in taking care of us without our conscious effort, and yet the belief in our free will finds much support in neuroscience labs. And we surmise correctly when we look at people and see in the data and interpretations of the data that we are both a part of the neural networks of our brains and yet strangely beyond those brains.

We will look through the brains and eyes of expert minds for help, from Nobel Prize winners such as Sir John Eccles of cerebellum fame, and Roger Sperry of split-brain fame. We will see famous writers who comment on our subject, such as T.S. Eliot, Marilynne Robinson, and the great Bard himself, William Shakespeare. All through the book, I will be referring to the work of eminent neuroscientists and philosophers, such as Wilder Penfield, V.S. Ramachandran, Christof Koch, Francis Crick, Patricia Churchland, and David Edelman, to name only a few. They are important voices on the main issue of the brain and personhood. Some of them I agree with, many I do not, but all are well worth reading and listening to. We need to think about this science before it is upon us, and it is upon us as I write. We undoubtedly need to reflect on our brains, ourselves, with the larger ethic and understanding of the Christian view on personhood in mind.

This book seeks to reclaim a sense of the sacred and the personal when examining the human brain. As the title suggests, this book will be about an almost inseparable relationship between our personhood and the neural activities and organizations of the brain. "Almost" is the key word. Evidence for our personhood exists in brain activity, in spite of a rigid reductionism in many neuroscience labs arguing to the contrary. The self-conscious mind peeks out at us from behind the veil of objective data and it tells of the richness and mystery of personhood. We will continue to see in this book with two eyes blending their observations: a scientific eye to examine the amazing data in neuroscience research, and an overarching eye to help us look with a sense of awe at the mystery of ourselves as persons.

****

"The brain's genius is its gift for reflection."
—Diane Ackerman (*An Alchemy of Mind*)

SOME BOOKS I THINK YOU WOULD LIKE

***An Alchemy of Mind: the Marvel and Mystery of the Brain*, by Diane Ackerman.** A great overview of the brain by a writer who knows how to write for all of us.

***The Man Who Mistook His Wife for a Hat*, by Oliver Sacks.** This is one of Oliver Sacks' first books, which established his reputation as seeing human persons in the midst of their neurological conditions. Informative, enjoyable, and compassionate.

CHAPTER 2

THE BALD BRAIN

An original way to see the brain and not lose sight of the person was creatively shown in a "hair-brained" lecture on the brain. Dr. Nancy Kanwisher in the Department of Brain and Cognitive sciences at MIT gave the lecture as she was cutting off her hair. As she lectured, Dr. Kanwisher sacrificed her hair with scissors and an electric razor, and then an artist came on stage and shaved Nancy's head with shaving cream and razor. While Dr. Kanwisher continued her lecture on the anatomy of the brain, the artist drew the brain's wrinkled cortex with a black marker onto her smooth, white scalp. The formerly gray haired neuroscientist had the artist draw on her bare scalp, illustrating the hills and valleys of the brain's wrinkled cortex. Then, as Dr. Kanwisher explained the functions of those parts, the artist colored in certain parts with red or green or blue markers to note the functions of those areas. Looking at the artwork on her professional head you did not fail to see Nancy, Dr. Kanwisher, the person, as a key part of the brain within. She was not just Wernicke's area or the motor homunculus in a textbook picture, nor a lifeless brain in a jar on the podium. But she was a talking, thinking, bald professor – a brain fused with personal life revealing this creative, intelligent woman.

A TITLE CAN SAY A LOT

Anatomy of the Soul

—Curt Thompson

BRAIN ON THE TABLE: THE ANATOMY OF MIND

"Swiftly the brain becomes an enchanted loom, where millions of flashing shuttles weave a dissolving pattern—always a meaningful one."

—Charles Sherrington

I am myself, and I am, in important ways, also my brain. My brain is awkwardly divisible into hundreds of separate parts, but I am to me more obviously a union of a thinking, feeling, and willing self. I am never just a network of neural cells, never a cerebellum, nor a limbic system. When touched with a scientist's electrical probes, or hit with an angry shout from behind, I immediately move and feel. I disappear every night in a slow wash of melatonin behind a veil of sleep. I awaken when I am gently prodded by more chemicals, and then more violently with my noisy alarm clock. Quickly I am back from my dreams, into the world of the awake, moving along with my busy thoughts and actions. I can see myself in a mirror, but I am definitely on the inside looking at me looking at the mirror image looking at me. I see myself from within my head, and with those mysterious mirror cells[1] inside my brain I can also see the inside selves of other people, even though I am not inside their heads.

The real you, likewise, is somehow experienced inside of your head, just behind the eyes, somewhere in there, in the brain in some way. But if we remove your brain and place it on the lab table and take a good look, there is nothing left of you to look at but a dead mass of cells. What creates the person in a live brain is what we want to keep in mind in this chapter. What we will see is that the mind is not mere matter, but it is somehow related to the massive numbers of neural connections between major brain areas. We will see that those unifying neural connections are not the creator of human consciousness. The physical unity of the brain's interconnections serve the purpose, amazingly enough, to lift your consciousness above the material realm and into an experience beyond the world. In a universe thought by some to be only matter, to see major areas of the brain devoted to placing you above the level of matter is something to take stock of.

The self without question begins in the brain. The soup and sparks of transmitters and neurons are important stuff of mind, but the question remains, are they all that I am? The human brain is made up of individual parts and pieces. But there is something unique in a brain when we see it act as a whole being. The brain functions like a busy newspaper office full of reporters having to write a single news article on the current political or athletic news. Some of the reporters are more respected or up-to-date on the subject at hand than others. The meeting hears from memories and facts in the important hippocampus, so prepared and proper. Then, sitting nearby is the amygdala, shouting out in the busy room with emotional comments on those remembered facts plus a few uncalled-for explosive swear words. There are more voices reporting in on other information on the subject at hand. Slowly, though, the group forms a consensus of blended sentences. Finally, one article is approved. Self feels like the chief editor who publishes the results of the committee work. I do not usually feel the parts and their blending. I feel in control of my thinking and my decisions most of the time. There is a "me" in my head. It seems that there is, somewhere, a self, a continuing person using feedback from important places everywhere on subjects of my choice. I am Humpty Dumpty back on the wall.

We must examine the human brain, pictured so well by neuroscience, and at the same time step back to see the whole person, just as we do when viewing a painting in an art museum. Claiming that standing with our noses on the canvas is the best way to see the painting misses the point that there is a larger picture, more real than the paint itself, to be seen. So, too, those of us who believe in God believe that there is a larger picture of human nature that can be rationally defended. Hopefully misunderstandings about strengths and weaknesses in science and faith can be avoided if those concerned are willing to work both in the realm of neuroscience, close to the painting, and are willing to keep a larger perspective on human nature in view while standing back from the painting. Neuroscience is revealing amazing things about human nature and at the same time making some out-of-bounds pronouncements about what is true about free will, purpose in life, and what happens when you die. The study of human nature is also being done in the humanities, philosophy, theology, and in our own experience. Therefore, it seems reasonable to keep in mind the complexities of the human brain along with what things we already know about the living, functioning whole person.

Anatomy of the Mind

You and your brain are one, a mysterious unity, a partnership of spirit and dust. We can see approximately 86 billion neurons interconnected with about one trillion, too-small-to-count glial cells squeezed into your skull. The bulk of these cells are in the outer part of the brain, the cerebral cortex, which is wrinkled like tree bark to allow more cells to be packed into the human skull. The surface of the cortex, if spread out, would be about the size of a large newspaper page, and about 2–3 millimeters thick. This outer cortex seems to handle the senses, movement, language, and cognitive functions. The wrinkled cortex is squeezed around the inner workings of the brain, which are also wedged in and made up of interconnecting neurons. These inside parts function together as an emotional-motivational network as well as areas related to movement. Again, these brain pieces have to be tightly packed because, if your head was any bigger, you would be top heavy and fall off bicycles, and of course, mothers would not want to give birth to such large-headed babies!

Most of the neurons in your cortex are small cells with an organized mess of tiny branches on them reaching out, trading electrical and chemical information to other neural cells in all directions. Some neurons are different in that they have a long cable-like trunk sticking out of one or both sides of the cell body. The trunk-like extension of these neurons, called an axon, carries a neuron's electrical information a longer distance than just next door before it passes the neuron's information along. All of these tiny cells are arranged into a weaving of electrical highways and chemical intersections that reveal a picture of the brain like a complex, crazy map of a city. This description is too simple because an even closer look at the cells of the brain will reveal at least a hundred different types of neurons along with liquid highways of important transmitter chemicals.

In the same way that the streets of a city are not the full nature of a city, we should be careful before we say the interconnected neurons and glial cells of the brain are you, the person. The brain's work is unified by these electrical and chemical pathways but not defined by them. A closer look reveals the chemical currents moving through intersections called synapses, and more extensive chemical pathways functioning as busy dimmer switches creating modulations or rhythms over the brain. Making neural maps of your brain or listing the functions of its parts does not exhaust what the brain is nor who you are as a person. What is important in the physical investigation of the brain is to listen to interpretations of brain activity that spring from our understanding of the whole of human experience and personhood. Just as with paintings, we need to step back a foot or two outside of neurophysiology and neurochemistry to see a more complete picture of what the brain's role is in the formation of a person.

We need to always look at the whole picture of brain function, the whole Dr. Nancy Kanwisher, the bold professor with the bald-headed brain. Then we will better see what the combined neural correlates of consciousness and other brain areas are contributing to the whole of personal experience. We need to study, but not get lost in, the tiny details of the brain, lest we believe by assumption that the whole person is never more than the pieces of brain tissue. We also cannot carelessly use the word "emergent" to refer to the more complex attributes of mind that we do not see in any brain parts, as if the immaterial mind and the richness of personhood just emerge without explanation from the combination of brain parts. We need to be willing to see the whole person as part of our data when we conduct our brain stimulations and scans if we are ever going to see how your brain activity relates to you. Anything less is avoiding the harder questions about an immaterial mind and the source of human personhood.

It seems clear that our consciousness relates to not just one source, but to the interaction of many different brain areas that connect neurally or chemically and function in harmony with one another. Brain areas such as the reticular activating system, the thalamus, the parietal lobe, the frontal lobe, and the claustrum are frequently mentioned. Examination of general anesthetics has provided clues as to what is involved in the brain when a person loses consciousness during surgery. The evidence seems to be that general anesthetics of most types seem to work by altering the interaction of major brain areas with each other. Consciousness, it seems, is not the work of one brain area, but the interaction of many brain areas together. Or, put another way, you are consciously present when the brain is operating as a whole and not just as so many parts. You are more than the sum of your brain parts. You are a whole brain/mind person, and in some ways even your body's actions affect your conscious mind.

Connected Brain Areas—Whole Person Functions

There is more to discover about the brain than simply a list of its brain areas and their functions. A comprehensive anatomy book could list hundreds of brain areas that have been investigated as to function, but the list would not bring us any closer to understanding the function of persons. The person is more than feeding, running, and breeding. We need to begin more holistically, with a view of ourselves as persons and what persons can do and think and feel. We need to reflect on what the human mind is and does when we examine the height and depth of persons and their accomplishments. The danger of looking just to areas of the brain for explanations of human personhood is that having too many puzzle pieces without a box-top picture for that puzzle tends to open up the door to anybody's picture of the human person.

It became popular in the early 1800s, to study phrenology, which was the correlating of the bumps on the head with personality traits. Phrenology has long been discredited, but some call the activity scans such as PET scans and fMRI the "new phrenology" because a quick look at scanning results could be used to show the activity of individual brain areas as responsible for complex human activities. The entire brain is active in complex ways for whatever a human being is doing, and it is not accurate to say that this spot of the brain does this and this other spot does that in terms of explaining human thoughts, emotions, and behaviors. The supposed discovery of God Spots in the brain as responsible for religious belief is perhaps the most discussed example of this way of thinking about the brain. Localization of function theories may reveal some things about the simpler aspects of a human being, such as walking and eating and seeing, but not the more complex aspects of persons, such as falling in love, choosing to become a missionary, or writing a poem. For those functions and many more we need to take the step-back view of larger, important brain areas interacting together, and see how human personhood relates to but is not explained by such whole brain activity.

Some of the history of looking for brain areas to correspond to human function includes the sad, well-known example of the railroad worker Phineas Gage and his accident with a flying crowbar that went zinging through his left, frontal lobe. On September 13, 1848, while Gage was tapping gun powder into a hole with his three-foot crowbar, the dynamite prematurely exploded, sending the crowbar through his brain. Phineas Gage lived 11 more years to tell the famous tale and, unfortunately, the exaggerated and many times inaccurate tales of Gage's behaviors after the accident became the definitions of frontal lobe function. He became the first accidental prefrontal lobotomy, if we could say it that way. Gage's personality was reported to have changed, and those changes became part of the clues to frontal lobe functions. It is difficult to tell exactly all the damage done to Phineas Gage's brain, since very few examinations or descriptions in any detail were written down. The accident was in the 1840s, after all. But the story helped reinforce the thinking about the frontal lobe's responsibility for various aspects of the human personality. In the first half of the twentieth century, thousands of surgical prefrontal lobotomies, cutting the frontal lobes free from major brain areas, were performed, such that by 1949 the Nobel Prize in Medicine was awarded to Dr. Egas Moniz for his pioneering work in the prefrontal lobotomy.

A surgical lobotomy is the cutting of the neural connections from deep inside of the brain to the frontal lobes. More exactly, the prefrontal cortex is separated from the rest of the brain without interrupting blood supply. The surgery today is not legal since we know its effects are much more devastating than merely removing one part of the brain of the depressed, anxious, or violent person. However, so strong was the view that the human personality was just made up of parts of the brain machine, that neuroscience moved rapidly in the direction of prefrontal lobotomies. Most of these surgeries were trans-orbital lobotomies, the inserting of a sharp, thin scalpel above the eye ball after a local anesthetic had been administered. Then the scalpel, sharpened on both sides, was moved left and right to do the cutting. A view of the whole person would have been much more cautious about entering into the human brain with a sharpened ice pick! But the zeitgeist of the time was the machine-of-many-parts view of the human brain and the person.

Major Brain Areas and Human Personhood

As we begin to look at important brain areas, it would be a mistake to think that each area does just only one thing, as opposed to considering the many connections each area makes with other brain areas. We cannot reject the idea that certain areas of the brain will specialize in or become involved as centers of visual or auditory activity, for example. But the overall functions of these many separate areas lie in their interaction with the circuitry of many other brain areas. Thus the brain operates as a whole unit in performing its many functions. With that in mind, we will now look at some brain areas of major importance. It is rare that large areas of your brain could be injured and you survive without trauma to you, the person. It is also never clear why some people who suffer brain damage do much better than others with similar damage. The answer is perhaps that the damage done to the interconnections between brain areas is different in each case. The problem is not so much what you lose when one brain part is damaged, but what happens when damage upsets the brain's ability to function as a whole. The major brain areas we will now briefly examine reveal interconnected functions that allow human mental existence to move outside of time, in symbols, in meaning, and in imagination. These functions of the whole brain help to stretch the human mind out of the material realm and into an existence that is mental and personal.

The outside of the brain is easily divided into four cortical lobes, with multiple functions assigned to each. It is important to look at three of these lobes, and other major areas, and ask what functions these brain areas orchestrate with their rich connections with the other areas of our cerebral cortex. The frontal, parietal, and the temporal lobes of the cortex, as well as the limbic system, the thalamus, and the corpus callosum (the connecting neurons between left and right hemispheres) seem important for contributing to what a person is as a conscious being. The amazing thing about what appears when these richly interconnected areas act together is the revealing of a human nature that is raised above the level of the material world. We exist in symbols, in imagination, in meaning, and outside of present tense. Yes, we still function in the sensory world of matter, but we mentally exist elsewhere. When we hit our thumb with a hammer, yes, we become one big, painful thumb for a few minutes. But most of the time we exist primarily in the nonmaterial world of our minds.

Human beings are in part material beings with material functions served by their brains, but they are also much more. Much of our consciousness is raised above and outside of the world of matter. Our richly interconnected, unified, thinking, feeling, artistic, romantic, theologizing human brains seem to be designed to allow us to stand up tall as beings above the world of matter, stretching us just high enough to see a massive universe, to sense the heights of love, to sense the presence of God, and to grapple with the ultimate questions of purpose, meaning, morality, and relationships. Let us now look at some important brain areas and what they contribute to you because of their interconnections with other major parts of the brain. The important point to make is that they function together to help detach our consciousness from the material realm.

Frontal Lobes—You Exist Above Time

The frontal lobes of the human brain make up the largest area of our cerebral cortex. The human frontal lobes include the motor cortex with obvious functions in controlling the body's muscles, Broca's area in the left hemisphere with its speech-production duties, and the prefrontal cortex and orbital frontal cortex, each with a variety of functions. We say that the frontal lobe functions include selective attention, problem solving, and other cognitive functions, including impulse and emotional control. We need to see that for these functions to be understood at all we must see them in light of their interconnections to other brain areas, and not get lost in the many smaller functions served by the brain.

When we examine the frontal lobes of the human brain with a view of what is distinctive about the human personality, we begin to understand the functions of its unique anatomy. The frontal lobes, with connections to the limbic system's hippocampus, make use of our working memory of the present, our long-term memory of the past, and our ability to construct a possible future (to *remember* the future), and thus to pull human consciousness out of present tense and into an awareness of the past, present, and future all at the same time. Additionally, connections to the parietal lobe's production of the boundaries of self, and the temporal lobe's symbolizing through language, allows us to exist mentally outside of the present moment. When a person has massive brain damage in the frontal lobe, or when there is sufficient disruption between the limbic system, with its memory functions, and the prefrontal cortex, that person may lose the ability to stretch his mind over time. Such a damaged individual can become a "two-minute person," who can no longer store long-term memories. In general, these damaged individuals can store and remember things for only a few minutes. They are living in permanent present tense. Oliver Sacks, in his book *Musicophilia,* describes cases of permanent present tense. One man has only 15 seconds of short-term memory. He will not remember anything new after 15 seconds have passed. His life is described as similar to always waking up. He is, in a sense, gone as a person if he cannot connect himself in the present to memories and to a meaningful future. We, as whole persons, were made to live above time, holding on to past, present, and future all at the same time. We easily gain insight into the meaning and purpose of events as we in the present, reflect on the past, and move with planning into the unknown future.

Parietal Lobes—You Exist in Meaning

The textbook descriptions of the parietal lobes of the human brain, located on both sides of the brain high above the ears, say that it performs two valuable functions for us. The parietal lobe integrates sensory information, primarily from vision, into a single perceptual experience. And it functions as the projection area, *i.e.*, the input area, for the body senses, thus helping locate the body and objects in space. In general, parietal lobe damage produces all manner of deficits to human functioning. In particular, there are problems with integrating inputs together into a larger, meaningful whole percept. The parietal lobe seems to bind combinations of stimuli into meaningful patterns. Our consciousness contains, in part, the meaning pulled from combining sensory information in the parietal lobe with frontal lobe contextualizations and temporal lobe language functions.

The parietal lobe and surrounding areas of the brain used to be called the "association cortex" because their apparent function was connecting sensory information together into assemblies that mean something. Meaningfulness comes from connectedness. An "A" grade in class is just a letter composed of three lines, and so is an "F," but when put the lines together in different ways, plus other cognitive input from language and frontal lobe functions, the three lines mean entirely different things. Damage to the parietal lobe and a person might look at a set of car keys and describe just their roundness or their length or the silver or shininess of the keys. Those descriptions are the stimuli of what you actually see. But what you perceive in your whole conscious experience are your car keys.

The parietal lobes, combined with inputs from the temporal lobe language-understanding areas and the prefrontal cortex help you put the stimuli together into a meaningful package. You do not see the colors and lines of your cat; you see a cute kitten. You do not see darkened brush strokes on a painting; you see sorrow. You do not hear individual notes of music; you hear "The Star Spangled Banner," or "Amazing Grace," and all the emotions of those songs are blended into a meaningful experience. This unity of thinking and being is not just putting things together, but developing the richness of meaning in the blending of those things. The meaning of something exists above the physical level of the object itself. Meaning depends on a connectedness, a unifying of our sensory world by parietal, frontal, and temporal lobe functions. With this larger function of meaning taken into account, it looks as if the human brain was more designed to read Shakespeare than to find food. With such a brain anything can become poetic as metaphor, narrative, and song play with images in the creative libraries of our minds.

Temporal Lobes—You Exist in Symbols

The temporal lobes have been given that name because they are located where the hair just above our ears first turns gray. Graying of the hair shows that time (tempus) has passed and we have aged. The temporal lobes, on both sides of the brain contain areas responsible for auditory (hearing) processing, some visual system processing, and language understanding in Wernicke's area on the left side of the brain in most people. The temporal lobes and their rich contact with both the frontal and parietal lobes help us blend meaning and thought in language to form symbolic experience.

The two speech centers, Wernicke's and Broca's areas and their connecting fibers, handle the most complicated functions of the human brain in language, which literally puts the person into symbolic, mental existence. The global destruction of Wernicke's area (well beyond its boundaries), which handles language understanding, produces a global aphasia and severely damages not only speech, signing, and writing, but also paralyzes the organization of thought itself. It is almost misleading to talk about the speech understanding center (Wernicke's area) and the speech production area (Broca's area) as completely separate areas with separate functions because thinking in symbols depends on the interconnections of these two areas with each other and with other areas of the brain. Global damage to these speech centers and related areas causes an emptiness of mind, that is, empty of connections or trains of thought. It is the speech center's connections with the frontal and parietal lobes that allow this important symbolic construction of the meaning of things into language.

Our mental life most of the time exists in symbols. One only has to think of the mind of the deaf and blind Helen Keller before she learned language. Helen's existence was largely the physical world around her. But persons with language do not exist just in bananas and bread meals. You exist in something out of this world—the symbol systems of community, relationships, and shared knowledge that language makes possible. Wernicke's area in the temporal lobe is the main area in a massive circuitry for language understanding. Without it we would not be able to think, except perhaps in artistic ways. Without this brain area we could not experience the minds of others, communicate our thoughts to others, organize and live in cultures and civilizations, pass on our knowledge to others, know the minds of even historical figures, and manipulate thoughts and ideas in our own minds. To say it like Peter Mark Roget of Roget's Thesaurus fame put it, "The use of language is not confined to its being the medium through which we communicate our ideas to one another; ...(it functions) as an instrument of thought; not being merely its vehicle, but giving it wings for flight."[2]

Corpus Callosum—You Exist in Imagination

The two hemispheres of your brain are connected by two hundred million neurons that allow the left and right sides of the brain to coordinate with each other and to share motor, sensory, and cognitive information, and thus share what is happening on each side of the brain with the other half. Such continuous sharing of massive amounts of information allows the rapid integration of complex information. This brings about a cognitive and bodily unity for complex tasks. A lack of whole or part of the corpus callosum produces deficits in movement, reading and speech, and some of the symptoms of autism. Knowledge of what it is like to have no or little connection between left and right hemispheres was shown in part in the Nobel Prize-winning split-brain surgeries done by Roger Sperry, but more clearly in the cases of infants with a failed development of a corpus callosum. The developmental condition is called the agenesis of the corpus callosum or ACC. In these cases the types of deficits seen vary because of timing and other brain anomalies accompanying the developmental problems.

Some of the social deficits from little or no corpus callosum include a lack of self-awareness, difficulties in taking the perspective of others, failing to interpret the social or verbal cues of others, and difficulties in abstract reasoning, sophisticated humor, and understanding the emotions of others. In the surgical splitting of the brains of epileptic patients, the person may develop hemispheric disconnection symptoms that last weeks or much longer, in spite of having grown up with their two brain hemispheres having been "connected" for their whole lives. The things that split-brain persons have learned before the surgery remain intact, but after surgery there is great difficulty in learning tasks that require interdependent movement of each hand, such as when playing a piano.

The split-brain surgeries of Roger Sperry showed that, in general, the two brain hemispheres tend to process information in different ways. The left hemisphere processes the world in sequential, rational, mathematical, and linguistic ways. The right hemisphere processes information in holistic, artistic, and emotional ways. With the corpus callosum intact, those 200 million fibers allow the sharing of each half of the brain's unique view of reality with the other. This mixing of the two hemispheres' different perspectives on reality, along with the work of additional cortical information, creates an imaginative product of the observed world that is not exactly the objective reality out there in the world. The left brain and right brain are processing the same information from the world in different ways. The reality you experience is a combination of these two viewpoints. The reality in your experience is slightly different than the objective world out there, and it is also slightly different from any other person's view of reality. You don't live in just the objective world, but you reside in a subjective world of your own construction, of your own imagination. This imaginative capability allows the great leaps of thinking and understanding that make human mind and culture what they are. Therefore, the left hemisphere of a split-brain patient's brain (which only controls the right hand), when isolated from seeing what the left hand of the patient is doing, will function as the interpreter for the patient's experience with a rich, imaginative explanation for the left hand's behavior. All of us live in that world of truth and imagination, which is constructed out of sensory information, attention, memory and expectations. We are living out of the physical world in a world of our own partial construction.

These major areas (the frontal, parietal, and temporal cortex, and the corpus callosum) by themselves work to serve different functions of the body. But when these main areas function together and with many other brain connections, they contribute to the higher aspects of human consciousness that reside outside of the material realm. We exist above time in our self-awareness, in meaning in all that is in life, in symbols that relate us to others and to the understanding of ourselves, and in imagination that helps us see and create new worlds out of our unique understanding of things. What these mental abilities mean for persons is that we can hold thoughts as symbols in our minds; we can manipulate and play with symbolic ideas over time; and we can create narratives with imaginative power, all in our mind's "eye." The world of the mind becomes more real every day in the world of words, and the invented world of our imaginations. That rich reality of human consciousness leads to libraries of symbols;

out-of-the-box thinking; shared ideas and culture; ultimate questions in philosophy and science; and the Word and words about God, morality, and immortality. The human brain/mind unveils the universe that has been prepared before us, and in which we can search out and find our meaning and our place. We are less the stimulus response robots of times past and more a race of story tellers living in larger narratives of discovery.

Einstein—A Person of Interest

One person who has helped humanity see the complexity and simplicity of the universe more than most others before him was Albert Einstein. Our look at Einstein's brain after his death showed that he had a very ordinary brain in most ways. And yet it was very different in other ways that spoke of its rich interconnectivity. Einstein's highly interconnected brain promoted a genius, who was like few other people in history. The year 2015 was the 100th anniversary of Albert Einstein's publication of his general theory of relativity that revealed that the force of gravity was not a force but a byproduct of a universe that was curved.

In 1905, the 26-year-old Einstein, working in a Swiss patent office, published four important papers in the scientific world, including special relativity and light quanta. The theory of relativity was expanded in 1915 to included gravity, which was confirmed by the bending of starlight by the sun during a total solar eclipse in 1919. He soon began to provide the steps leading to the development of quantum theory, which led to his Nobel Prize in Physics in 1922. Einstein thus changed the face of Physics forever. Nothing like that had happened since the time of Isaac Newton. And just recently in February 2016 we have heard that Einstein's theory of general relativity has been confirmed again in that gravity waves have been measured for the first time.

The very physics of the great Isaac Newton had been overturned and our whole concept of space and time was changed by the brilliant mind of Albert Einstein. Einstein became the very image of the thinker and the genius, with his shy, withdrawn, quiet demeanor, and his theories that few understood. And, yet, he was our teacher with clear explanations and examples and thought experiments that broke the barriers from the genius to the common citizen. He was such a recognizable person for all of his life that George Lucas had an artist base the face of the Jedi Master, Yoda, partly on Einstein's face, particularly Einstein's droopy, solemn eyes for the wise Jedi Master in the Star Wars films.

Einstein was not just a brilliant, isolated thinker, but he was an ordinary man too. He was a young father and husband with a job in the Swiss patent office. He quickly became famous, but fame separated him from his wife and family, leading to his isolation as a professor in Berlin. Years up to 1915 saw the completion of his famous general theory of relativity and increased tensions with his wife and two boys. As Einstein interacted with other great minds in the world of science, he experienced both support and tension.

Over the years he would also experience war, antisemitism, and his search for meaning in life. Playing the violin brought him some mental relief. Brains do not develop nor work in a human vacuum, and Einstein's did not either. He was a person rich in thought and emotion in all areas of a human life from personal love to politics and world peace.

Though Einstein died in 1955 and was cremated, we do have his brain and can explore with science why he might have been so brilliant and could see things hidden in the universe that no one else could see. When Einstein died at the age of 76 in Princeton Hospital, Dr. Thomas Harvey, without the family's permission, removed Einstein's brain and preserved it. He later received permission from Einstein's son to study the brain, but it was only to be used for scientific purposes. What was discovered by experts examining pictures and pieces of the brain of Einstein was that it was difficult to pin Einstein's expertise down to any major brain differences. There are good reports that Einstein's brain had up to twice as many glial cells as the normal brain, which may be important because of what we know about the importance of connections in the brain served by glial cells and their roles discovered recently in learning and memory. Einstein's brain also possessed a smaller fissure on both sides of his brain separating the parietal and temporal lobes, leaving two major lobes of the brain more connected, including an area normally thought of as relating to mathematics. Einstein's brain had a larger than normal corpus callosum, which again speaks of an exaggerated connectivity not found in many human brains. Einstein's brain, with a larger corpus callosum connecting the left and right hemispheres, probably had many more brain areas with different functions working together than the average person. Einstein's brain also showed an enlargement of an area of the brain considered responsible for moving the fingers when playing the violin. But nothing else about his brain suggests anything out of the ordinary. Einstein's brain was a bit smaller than the normal brain and that did not seem to matter one neural bit.

The Connectome Project

Given the importance of connections between the major areas of the human brain and its various networks it seems reasonable for neuroscience to now undertake what is called the Connectome Project, to map the human brain and all of its connections. With a complete neural map it is thought that we would better understand how the brain and its individual networks are wired and how they work. To that end, as well as the knowledge that can be gained from a detailed knowledge of the brain's networks, mapping of the brain has begun. Recently, the government announced a three-billion dollar, 10-year scientific campaign to map the human brain. What the Human Genome Project did for unraveling the story of genetics, the Connectome Project, will be doing for the difficult mystery of the brain. The enormous project will initially involve two groups of schools united in the research effort. One group will be led by the University of

Washington, St. Louis and the University of Minnesota and the second consortium of schools will be led by Harvard, Massachusetts General Hospital and the University of California, Los Angeles.

To map all the interconnections being made in the human brain will not be possible any time soon. But neuroscience can begin with a map of the connections between the major brain areas, which will make medical assistance to the victims of stroke, Alzheimer's, and Parkinson's easier to develop. Neuroscience has made a map of the "brain" of the worm, known as Caenorhabditis elegans (*C. elegans* for short; aren't we glad for shortened biological names?). This worm is only one millimeter long, about a tenth of an inch, and has a mere 302 neurons. We have known the map of this little worm for 25 years and given that there are a possible 100 different neurons involved in its connectome, it is incredibly complicated for its size. That number of types of neurons is also far too many types of neurons needed to explain this simple worm's behavior. Recently a team of researchers from Princeton University were able to image its tiny brain of 302 neurons while *C. elegans* continued to swim. The neuroscientists were able to correlate multiple neurons to the worm's movement behaviors of crawling backward and forward, and turning left and right. But we are still a long way from understanding this worm's "mind." Mapping the connectome for the human brain's 86 billion neurons, trillion glial cells, and untold trillions of interconnections of neural and glial pathways is mind boggling to say the least.

But in spite of the difficulties, the research to map the human brain is proceeding in earnest. Using light microscopy to image tiny brain slices down to a few hundred nanometers (billionths of a meter) is sufficient to image structures as small as neurons and dendrites, but not their synapses or points of connection. It was thought that neuron parts being in the general proximity of each other could be counted as actual synaptic connections, but that has not proven to be the case. More recently electron microscopy or EM has been used, that is, using electrons instead of photons of light for the viewing of tiny brain pieces and thus seeing much more detail. But imaging with this tool is more expensive and difficult to use. Therefore, light microscopy is still used with the limitation that what we are seeing as connections may not be true connectivity at all. Other tools for this mapping research are improving as well, such as diamond knives capable of making 29 nanometer-wide (billionths of a meter) slices of brain, and improved pattern-recognition software capable of recognizing glial astrocytes and synaptic vesicles carrying transmitter chemicals across the synapse.

With all that technology and precision, and many labs operating, it would still take many months to map a three-dimensional piece of brain one-trillionth the size of the human brain. And the data storage requirements for the human connectome right now represent a large fraction of the storage capacity on earth at the present time! Even so, with improvements, it is hoped to acquire a complete map of 1 cubic millimeter of the human brain in five years. That is a bit disappointing, though, in that there is so much more brain to map.

Even a complete map of the human brain will not tell us how the brain works. The simple connections of neurons in a network with either excitatory or inhibitory connections in the map is not following the path of perhaps a hundred different other types of neurons that differ in their surface receptors, neurotransmitters, electrical characteristics, or gene expression patterns. And if the ideas of astrophysicist Roger Penrose are any indication, we may have to extend that Connectome map to the cable network inside of every neuron and eventually to the sub atomic activity within that network. Like the Human Genome Project, the Connectome Project may only be scratching the surface of the most complicated piece of material reality we will ever encounter.

There are also other problems with maps of the human brain. If mapped, the human brain would be a street map in three dimensions of a million cities and trillions of highway intersections. And those difficulties are only compounded when we consider that the brain is not just made of neuron pathways. The brain also has chemical highways in its transmitter chemicals and volume transmission systems. And any brain we map in real life is small, cramped, and composed of wet tissue ready to fall apart when touched. Furthermore, the street map of a city is not the city. A map is static. You may be seeing Boston streets, but not what those streets are like at midnight or noon, during a traffic jam or a snowstorm. But any attempt at a map is a start, say many.

Closing Thoughts

We need to be reminded that the map of Boston is not Boston. It is only a map. And thus, in the Connectome Project the nagging question about the human person and the brain remains. What will be missed in the big map of the brain is the living person, the mysterious matter of you. You are a mysterious being of spirit fused with the matter of brain, and an accurate explanation defies our labels of dualism, monism, nonreductive materialism, interacting dualism, duality without dualism, and more. No map, no philosophical label does justice to you, or solves the ever-deeper mystery of you. You are a mystery of two parts made one, matter and nonmatter wedded together until death do they part, and wedded again once more in the resurrection.

When we human beings see and ponder the massive and tiny pieces of the universe, from its galaxies to its sub-atomic particles, we notice that human beings seem to sit halfway between these two extremes, almost in a privileged seat, to observe much of all that is. Our minds/brains sit in the middle of our skulls, balanced neatly over the center of our shoulders, as we think about the universe's parts both big and small. We also seem to be sitting in the middle of time, between the long ago and the yet to be. All these parts of the universe are slowly giving up to our attempts to understand them, even though we have a mere three score and ten years to study and reflect, and a mere three pounds of wet brain with which to do our wondering. When I see that we mind/brains are sitting in the middle of a universe of matter, attempting to know the complexity of genetics and galaxies, let alone the reason and purpose of it all, I am amazed. I see that we humans are able to peer inside our own skulls and see the greatest wonder of them all, ourselves as persons, spirit fused with matter. It is interesting that we persons of mind/matter seem to be the only material realities around spending time looking upward and inward.

Now I am thinking about Dr. Nancy Kanwisher again with her bald head and her painted brain for all to see. When researchers reach inside the skull to reveal what is there, they too easily forget what is wholly and holy there. They might just count cells, follow pathways, and analyze brain scans. I suppose that is to be done, but maybe it could be done without the grand pronouncements of reductionism, that only matter was found in the skull. "There is no person in the brain," is reported, "only three pounds of water and fat." But when we see the brain painted on a woman's scalp, we get the point. Dr. Nancy Kanwisher is there, outside and inside. The brain is there inside, joined, fused, united with, and one with her. Even a "Ghost in the Machine" model is better than the machine-only view in the face of the magnificence of what human beings can be elevated to become.

****

"... there is a definite tendency to avoid explanations of human behavior which appeal to the conscious decisions of persons in favor of almost any non-personal factors."
—C. S. Evans (*Preserving the Person*)

SOME BOOKS I THINK YOU WOULD LIKE

***Mapping the Mind*, by Rita Carter.** This is a textbook of sorts, but has a form and a writing style that captures your interest in the brain and mind, and some of the issues of the most complicated physical structure that we know about in the universe.

***The Brain: The Story of You*, by David Eagleman.** Neuroscientist David Eagleman journeys into the human brain and ourselves in this book, which accompanied the PBS series, *The Brain*. He simplifies the important facts and issues that surround the human being and the brain's relationship to emotions, criminal justice, robotics, the search for immortality, and more.

CHAPTER 3

THE LIQUID BRAIN

The ancient Egyptians, who have been recognized for their early knowledge in scientific and medical areas, thought so little about the human brain that they threw it away when mummifying the deceased. The other vital organs they saved for the afterlife, but that would be little compensation to the royal person who was traveling through eternity without a brain. The surgeons responsible for preparing the dead for eternal life actually turned the brain into a soup with lumps of brain instead of vegetables. They stuck a pointed tool up the nose, through the skull, and moved it around until the brain drained out of the nose. Some of the tools used by the Egyptians are on display in the British Museum in London. One tool cracked the bones behind the nose, and another slowly drained the brain. It was little loss, thought the Egyptians, who then filled the cranium with linen or straw. Dorothy's straw man in *The Wizard of Oz* knew better as he sang, "If I only had a brain!"

A TITLE CAN SAY A LOT

Wet Mind

—Stephen Kosslyn

RIVERS OF THE MIND: SHAPING THE SELF

"Is it not strange that sheep's guts should hale souls out of men's bodies?"

—Shakespeare's Benedick
(*Much Ado About Nothing*)

Every function of the body, including the brain, runs on water. Water is needed for the production of the brain's neurotransmitters and hormones, without which we would be back to the dust of Genesis. We do not forget to leave a water bowl for the dog when we leave the house for the day. However, we do tend to ignore the need to take frequent drinks of water for ourselves, especially as we age. Water helps provide nutrients to the brain and remove toxins. Prolonged dehydration causes our brain cells to shrink. Water is what we look for from so far away when we search with our telescopes for the possibility of life on other planets. So, too, it is in the fluids of the brain that we may find our next clues about the human person, not in the more rigid neural networks, but in the rhythms of the brain brought about by transmitter chemicals.

Our brains are largely water, specifically about 78% water, 12% fat, and a few other odds and ends. Water does not bring intelligence, and adding fat to the mix does not make it any more likely that your species could have the brains to send men to the moon and back. If we played the Egyptian burial game with all those moon scientists, we could drain every brain out through the noses of those now-deceased scientists and find nothing there but water. The human brain may appear solid; you could hold one in your hands, but it is not just a set of wires or axons to follow and create a map to see where everything goes. It may be easier to think of electrical wires in our skulls when we think of our intelligent species, but that view of us as computer robots may have gotten us off on the wrong connectome track to the understanding of our personal thinking and feeling selves. On the other hand, to correlate our thoughts and feelings with the liquid waves of the brain, while it may be more difficult to imagine, may be far more accurate. Our minds seem to be more rhythm and less mass.

The Limits of the Connectome Map

The Connectome Project is an exciting development in the study of the brain, but it is certainly not the end of the matter of finding out how the brain works or what the total physical platform for human personhood is. The Connectome map will not be the end of the human brain story. The Project will not give us a complete picture or the human brain, and in fact it may be a misleading picture of what the brain and personhood are all about. We are neither computers nor robots. To believe that we are machines will only leave us with a partial picture of the total of what brain activity is all about and what it contributes to personhood.

The Connectome Project will give us pictures of the larger neural pathways and their major intersections, and some fragmentary mapping of the simpler neural networks. That is fine if you believe that is all there is to a human person. The emphasis, however, is on being able to plot the highways, or a road map, at one moment in the brain life of a person. The assumption is always, of course, that there is nothing else functioning in your brain but fatty neural membranes with their leaking ions that make the roads and the network programs from which, supposedly, emerges you and your personality. Take a picture and there you are, a set of genetic programs, learned connections, and memories. How the immaterial mind and self emerges, we do not know. What we do know, though, is that like a living city, a living person is more than just the map of the brain. There is dynamic activity in the people who travel the city streets and what changes every second, every day, all year long, to make the city come alive. You are not just your static biology, a snapshot of your muscles and bones, and your neurons and skull. Those things help make you what you are, but such a view is ignoring the subjective you inside, which is a dynamic, changing, thinking, feeling you. The longer we ignore the person of you, the longer will be our journey to understand the human person beyond the superhighways of the brain. A look at some of the chemical activities of these neural highways will take us one step further in seeing material activities that run parallel to the flow of the mental and emotional inner life of a person.

Chemical Highways

The chemistry of the brain is not a soup of many ingredients from an Egyptian cookbook. The brain's chemistry acts like another set of brain pathways besides the connectome highway, more flexible and constantly changing. Transmitter chemical streets function, not like the semi-rigid neural pathways that make up the brain with their fixed patterns of action potentials and the hardened circuits of memory, but more of the artistic movements of a song or a dance. The dynamic functions of the brain are not fixed in paths, not set down in Roman roads of stone neurons, but are arising from a liquid, present-tense splash of transmitter chemicals, and constantly changing waves and whorls that form the rhythms of brain activity.[1] This dynamic wave form, or modulation, of activity may be found even down to the rhythms of the quantum world within the synapse, inside of the cable networks of the neuron, and in the slower rhythms of glial cell activity. Just like the rise and fall of the human voice may carry a message of meaning and emotion, so too, the fluctuating, wave-like motion of these chemical rivers are capable of revealing some of the complex patterns of mind.

The liquid, rhythmic, modulating picture of brain chemistry adds to the theories of how we as persons might be related to the world of brain matter. The rigid neural roadways are what might be the structure of a human brain, but it is the addition of the fluidity of change and what that does to thought and action, that contribute to us as unique individuals, all seven-and-a-half billion of us alive today. The connectome neural map might be thought of as an orchestra of musical instruments just sitting there. The modulatory chemical transmitters of the brain, which we will soon talk about, seem to be providing the score that the instruments are playing. With that example, the person becomes the music, the symphony, and, thus, we are more like musical chords playing a song than the machine language of ones and zeroes.

The brain's neurons fire electrically as a result of the brain's many stimulus inputs, motor outputs, and the constant use of memory and learning formations. This constant neural activity of action potentials is moved with a rhythm or modulation by long-range volume-transmission chemical highway systems. You could say that a song is constantly being played on the brain, turning the activity of billions of individual neurons into a unified concert of rhythm. We do know that our chemical transmitters, like so many dimmer switches, modulate the activity of neurons. Our chemistry plays music on networks of neurons, and thus brain areas are the instruments to play, the action potentials of neurons are the notes, and we can be our own conductors if we practice and learn our music lessons.

We will first look at the chemical pathways of the brain, which are not at all like the superhighways of the neurons to be mapped by the Connectome Project, but more like the rivers, tributaries, and creeks in the rural countryside. The transmitters are important for many things, but are very important for what we call our personality. Certain transmitters have huge impacts on the functions of the cerebral cortex, limbic system, brain stem, and the cerebellum. All of these areas are related to our humanness, parts bound together toward a personal awareness. These chemical modulators help to bind the activities of movement, thinking, and feeling into a whole-person rhythm.

Transmitter Chemicals—Local Effects between Two Neurons

Your brain is awash in chemical transmitters numbering dozens of types and functions. Usually, each neuron makes contact with 5 to 10 thousand other neurons across a tiny space called the synapse. The electrical impulses along the neural trunk have to pass the synapse, and to do that they need chemical transmitters that will transmit the electrical message to the other side of the synapse. Each neuron usually has from two to six different transmitter chemicals, which are responsible for another neuron across the synaptic space to electrically fire or not. If the neurons in our brains had only to send or not send on an electrical current, at most you would need only two transmitter chemicals—to go and not go. But we know there are probably at least 100, and maybe more, of these chemical transmitters. These chemical transmitters also have a two-way interaction with glial cells surrounding the neurons. Much more is going on in addition to the simple conducting of electrical potentials in neurons.

The brain is more complicated than just groups of neurons speeding electrical messages through synaptic connections, because much is happening chemically at the level of the individual synapse. These chemical transmitters begin the tiny leaks that can start or stop the electrical surge through the next neuron. These transmitter activities produce what are called ionotropic effects that relate to whether other neurons will continue the electrical signals being passed. Other effects of transmitters are called metabotropic effects, which will change the following neurons in ways beneficial to learning and memory. And, finally, some effects of transmitter chemicals are called modulatory effects and do not usually make neurons fire, but make nearby groups of neurons raise and lower their levels of firing. These modulatory transmitters affect a large number of neurons further removed from the releasing neuron.

Transmitter Chemicals—Local and Long Range Effects on the Brain

Most of what we have learned about transmitter chemicals concerns their rapid and precise synaptic activity. This is synaptic transmission and involves the release of transmitter chemicals from the presynaptic membrane of the neuron, with the effect of either excitation or inhibition of the following neuron. But transmitter chemicals may also have effects that are slower and over a broader area. This second type is called volume transmission and is more suited to the modulatory functions of transmitters. Transmitter chemicals can spread outside of the synaptic space by what is known as local volume transmission. This happens locally, but away from the tiny synapse. These effects come from transmitter chemicals in the neuron that are released more slowly from the neuron and usually not into the synaptic space. They diffuse in the general area around the neuron and modulate or control the firing of nearby neurons.

Transmitter chemicals can also have long-distance volume-transmission effects arising from a small number of neurons in the brain stem, which give rise to large systems of chemical communication that are interconnected. The transmitters most often mentioned in these volume transmission systems in the brain stem are DA (dopamine), 5HT (serotonin), NE (norepinephrine), and ACh (acetylcholine). A key function of these long distance volume-transmission systems in the brain is modulatory, that is, small networks of neurons in the brain stem, numbering only in the tens of thousands of neurons, begin the release of transmitter chemicals that have modulatory effects over extensive areas of the brain. These modulation transmission systems act like liquid highways of rhythm. They are small numbers of neurons affecting huge brain areas, and thus are labeled volume transmission.

The primary effects of these volume-transmission systems are modulatory, raising and lowing the level of activity in distant areas of the brain. This is what we mean by modulation. Like AM (amplitude modulation) or FM (frequency modulation) radio, these systems are capable of carrying messages. They raise and lower the level of firing, like constantly changing dimmer switches. You would not want to play with your light dimmer switches in your home. You simply raise or lower the lights in the room and then stop. But these dimmer switches in the brain are constantly changing and presumably correlate continuously with the rhythm of mood and thought in your mind. The four "dimmer switches" of DA, 5HT, NE, and ACh are important because slight changes in the activities of these transmitters can drastically affect the human personality.

If you throw a tennis ball into your swimming pool, the ripples from the ball entering the water would carry some information about the size, shape, speed and weight of the ball. And if you jumped into the pool, your ripples and waves would say something about your size and weight too. Information can be carried by rhythms in any medium. So, the modulation of waves over the brain carry some information about what is going on in the brain, that is, what you are seeing, hearing, feeling, thinking, doing, and more. Each of the four volume transmission systems can modulate different and overlapping brain areas in ways that say something about you, and the resulting combination carries a more complete picture of your mind at any one moment. If both you and the tennis ball hit the pool at the same time, all that information in the resulting waves is telling us something about what hit the water.

The brain stem volume-transmission networks act as dimmer switches that turn the modules of our brain activity up and down, creating neural patterns that in particular spell out our moods. Emotions are similar to the storms in our weather. They arise quickly, are hard to control, and are usually felt, physically and mentally. Emotions exist as a rhythm in the ups and downs of our feelings. Disrupting the rhythms of the brain may very well be a part of the emotional flow of depression and bipolar disorders, not the entire cause, but a contributing factor. Rhythms in the brain and body may just elevate or they may just as easily tear down the structure of personality. Understanding the nature of emotions or the whole person is obviously going to be more complicated than the connectome mapping. The investigation of the transmitter effects at each synapse, throughout the ventricular CSF space, and lastly in the volume transmission systems from the brain stem is daunting.

Liquid Personalities—Transmitter Chemicals NE, 5HT, DA, ACh

The fact that these four modulating transmission systems have so much to do with your personality tell us that we have located a more important flow of information concerning your personhood than any brain area identified as a neural correlate of consciousness. You are not the liquid transmitter, but you are somehow related to the rhythm, or the message being produced by volume transmission modulation. You can modulate finger tapping into an SOS message of help—three long, three short, three long. So, too, very complicated messages can be produced by waves of chemical and electrical rhythms in the brain.

For personality problems, such as depression, and some types of schizophrenia, the problem in the brain is not an abnormality of brain structure but one of brain chemistry, or both. Levels of the transmitters, too high or too low, often bring on changes in a person's ability to handle changes in life's circumstances. The importance of this for purposes of my discussion in this book is that the relationship between the matter of the brain and you is related, not just to neural firing, but to the rhythms of neural firing of the brain. You, your personality, seems more related to the modulated messages revealed by these chemicals. It is interesting that modulated activity can contain a message and that you might be described more in the message of the brain than just the brain matter itself. A letter you write to your mother is more than paper and ink. There is meaning present in the arrangement of the physical materials. So, too, your brain may be producing a rhythm, but the message is somehow related to who you are. Your personality might be an overlapping message of at least four, maybe more, volume-transmission systems that create one unified message with meaning, which somehow speaks of you, or perhaps is you. In a healthy person these four volume-transmission systems are all functioning normally and uniquely. When the brain that supports these systems is damaged, or when other chemical or learning factors due to stress or learning interfere with their normal, ideal operation, then the personality is hurt and even changed over time.

NE—(also referred to as noradrenalin) The Noradrenergic Pathway, using the transmitter norepinephrine, begins in the brain stem area called the locus coeruleus. This tiny area is on both sides of the brain stem, each side having approximately 25,000 neurons, a small number given the number of neurons in the brain. While the average neuron in the brain makes synaptic contact with 5,000 to 10,000 other neurons, an incredible amount in itself, the locus coeruleus neurons make contact of one to 250,000 neurons. These contacts literally encircle the brain and affect almost every brain area, including the brain stem and cerebellum. One way that this transmission system acts is to respond to something that is startling, such as a threat or a loud sound. Thresholds to firing are lowered and more neural firing occurs in areas that can respond to such stimuli. Disturbances in this pathway eventually relate to affective disorders like depression. When there is too little NE, depression in the personality may result.

5HT—The Serotonergic Pathway uses the transmitter serotonin, shortened to its chemical name 5HT (5-hydroxietriptomine). These neurons normally govern mood, and they originate in the brain stem in the area called the raphe nuclei, 300,000 neurons on each side. They make a whopping one to 500,000 connections each as they circle and affect the entire brain. When serotonin decreases, depression may result. This fact has led to the development of the drug Prozac, which is a second-generation antidepressant, an SSRI, a selective serotonin reuptake inhibitor. Essentially this means that when 5HT is released into the synapse, most of it is not taken back into the presynaptic, releasing side of a neuron, and therefore remains in operation longer, reducing depression symptoms. It is good to say at this point that nothing relating to the human personality is so simple that one pill does the job. Serotonin affects many other systems in the brain and may only be affecting depression indirectly.

DA—the transmitter Dopamine, usually associated with a reward system, is found in two separate brain locations, the ventral tegmental area, and the substantia nigra, which innervate motor, emotional, and cognitive functions. Two much activity in the frontal lobes because of dopamine-producing neurons pushes the vulnerable person towards schizophrenic symptoms. Too little DA being released in the substantia nigra of the brain stem relates strongly to Parkinson's disease symptoms. In Parkinson's disease the substantia nigra dopamine-releasing cells are slowly dying and the person is becoming progressively less and less able to control muscle movement in addition to other Parkinson's symptoms. In addition, DA activity in the limbic system seems to be important for motivation because of the experience of pleasure and reward. Cocaine is well known for enhancing dopamine production.

ACh—Acetylcholine is released from the basal nuclei of Meynert and the medial septal nuclei, and the pontomessencephalotegmental complex (Wow! Just think "originating in the brain stem" and you are close enough). Loss of this transmitter relates to learning, memory, sleeping, and Alzheimer's disease. We have so much more to learn about this massive transmitter chemical system and its role in matters of the brain and mind.

These four transmitter chemicals are referred to as volume transmission because of the huge extent of their influence. They are also modulators, that is, they do not function to make neurons fire, but they raise and lower the levels of firing of neurons and neural systems. They are the dimmer switches in constant motion, raising and lowering the general firing activity of the brain's neurons like a musical conductor with his baton producing a song or a concert out of the orchestra's instruments. These neural systems relate to you and your personality as they affect the bulk of the brain's activity. If they are damaged or changed, your personality suffers, as in depression or schizophrenia. When they run smoothly, you are you. You really are a song, or a symphony. You are four symphonies all playing at once. The four different systems are functioning not like individual notes but more like the flow of musical chords in a song. You are a message, a complicated message at that—not a rigid machine, but a flowing piece of art and music.

These four modulatory transmission systems exercise a creative control over emotions. Human emotions are not sensory, such as hearing a sound or tasting bread, but are complex feelings in the mind. They are cognitive and meaningful ideas, tied up with complex feelings. They serve the purpose of enjoyment or pain. They are the hidden power behind motivation, for without emotional preference one way or another, we tend to not do anything. It is usually fear that makes us run from the bear. It is love that makes us seek out another. Feelings make us tend one way or another, to go here now or to put off an appointment. It is with emotions that we decide whom we favor for president, or whether we should vote at all. We seek positive emotions most of the time, and when these emotions have cognitive support, we more actively pursue our goals.

I believe it is reasonable to say there is a two-way interaction between you and this brain symphony. These chemicals affect you for better or worse, and you can affect the rhythm and activity of these chemicals. This is, of course, the mind/body problem, which neuroscience often solves by saying you are equal to the brain's activities. Obviously, in this book I am looking at evidence, in the brain or in real persons, that preserves our concept of you as a person with mind and will and an existence beyond the death of the brain. We are not simple computers or brain maps. We are beings of more than matter. Obviously, this essence of a human being makes our search for the nature of the person in a complex brain a difficult search to say the least. Every neuroscientist should be cautious before proclaiming that understanding the brain is just around the corner, or expecting that easy solutions to mental and emotional problems will be easily solved with the latest drugs or technological advances.

If we believed that the purpose of our brains and consciousness was to accurately report on the world around us, then we would begin to think that the brain does not do its job very well. The truth is the brain does not report what the world is, but it reports it as we are, as what our expectations, motivations, state of mind, and biases are. The reports we get in our conscious experience are accurate enough to get us through life, but the brain also seems to make us into artistic beings with emotion and feelings that shape our thinking. To separate reason from emotion produces abnormalities in behavior in human beings. To develop both sides of ourselves produces the artistic nature in us and our very conscious selves becomes an art piece. We are more like a painting, a story, or a song than an unconscious computer code. We are "poetry in motion," as a popular song once said.

It is interesting that when the brain is scanned with the new two-photon microscopy scanning technique, the resulting images of the dynamic networks of intact brains, what appears in our pictures has been described as a grand symphony of brain activity.[2] Since each neuron can be said to produce a unique melody with its electrical signals, groups of neurons in complicated networks become as orchestras, playing symphonies as a whole unit.

Human Consciousness Is Similar to the Rhythm of Music

Human beings seem to have rhythm as a part of their lives, which we might have expected, with music playing a major role in brain activity. Human beings seem to have an innate appreciation for music, which is shaped by the culture in which we live and mature. Babies in the womb will respond to music, and after birth show positive responses to rhythm, melody, and the harmony of music. While the right hemisphere of the brain shows increased activities for music, it is really both hemispheres interconnecting during listening to or performing music, which describes the process. This is exactly what we would expect to come out of an interconnected brain. Both hemispheres and frontal lobe connections are needed for the complete perception of and appreciation of music. Perhaps this is why we find music useful and even necessary, as it enhances learning, lowers blood pressure, relieves pain, improves athletic performance, and develops better emotional responses.

A recent study connected fMRI results from brain areas activated by mathematical beauty to the same brain regions related to great art and music.[3] The common areas were in the medial orbital-frontal cortex, just behind the forehead. If brain rhythms were helping connect broader ideas of beauty, we would expect there to be a common experience called beauty in a human being. Mathematicians have often compared their experience of beauty in math, for example in the Pythagorean identity, to the beauty experience in paintings or music. Music activating this area also prompts memories for the time or event or person with whom the music was associated, showing a combination effect similar to what the binding problem is trying to solve.

Oliver Sacks in his book *Musicophilia* suggests that music has an area of representation related to rational speech, and another part related to the emotional experience of music. This separate part for the emotional experience of music may explain some of the effectiveness of music therapy for children, why children's singing helps improve speech by improving motor abilities. Most adults will attest to the power of music to move one emotionally and rhythmically, when one is depressed or when one has to engage in athletic competition. The healing power of music includes assisting patients with brain disorders to make progress toward the recovery of language, hearing, motion, and emotion. The human brain's functions for most adults can proceed in the emotional bath of musical sound. It is music's effect on the activity and distribution of neurochemicals in the brain, and thus the rhythms of the brain, that may aid positive emotions and even a certain amount of healing. Music can also deal with issues of timing, rhythm, and coordination that help patients with strokes or Parkinson's disease.

Some research from the Department of Homeland Security's Science and Technology Directorate involves a form of training called neuro-training. This training involves the collection of the brain's own "music," *i.e.* its waves of tempo and amplitude that change with your mood and thoughts as well as some individual features of your brain. The music created from this information is collected and used to produce two short piano compositions, one for relaxation, and one for alertness. These compositions are then played back to the first responders to calm their anxieties of the job, or to business employees to improve productivity at work. The thought is to use your body's natural musical rhythms to increase energy or to reduce stress. More research needs to be done on these ideas, but it reminds me of Snow White's seven dwarves who sang "Whistle while you work" as they labored. Music and rhythm are part and parcel of who we are as persons. We are not the chemistry of the transmitters, nor the overall brain waves they create. We are more like the message in the combination of all the brain waves. We are the musical chords of volume transmission systems that combine to make our individual "musical, mental self."

Music, rhythm, gesture, emotions, and the feeling of self in space is you when you dance. The subcortical circuits in the cerebellum provide the connections between feedback coming from your moving muscles and your kinesthetic sense, with the basal ganglia at the center of the brain, and the cerebral cortex itself with its control over the learned movements in dance. Neuroscientists have put dancers in PET scanners and fMRI's to see where all this occurred.[4] The dancers were flat on their backs in brain activity scanners, heads not moving, but they had the freedom to move their feet in dance rhythm to the music they heard. At other times the dancers merely thought about dancing to the music. Sometimes they could flex their leg muscles, but not move their legs. We can see your brain in action as you even think of dancing. Who says that modern scanning techniques cannot be fun?

William Shakespeare—A Person of Interest

We marvel at the inexplicable genius of a music prodigy such as Wolfgang Amadeus Mozart. He was composing music before the age of six and went on to compose amazing symphonies and sonatas, and it is rare that other musicians have been found to be his equal. But we do not tend to think of someone who was an expert with words in the same breath as a genius in music. But the use of language is rhythmic and poetic just like music is. We are not geniuses like Shakespeare in the use of words, but one of the most obvious characteristics of a human being is the use of language, which is mental existence in symbol systems. It is such a linguistic system that gives shape to our personalities and affects our perceptions of the world.

We actually know so little about William Shakespeare, even his birthdate is unknown. We have only about one hundred documents that are from the bard and his immediate family, and most of those are just deeds and court records. As Bill Bryson puts it, such things are "inevitably bloodless." They tell us little about the man. But we do see Shakespeare in the flesh, in the mental flesh and paper flesh of his characters and their stories.

The genius of William Shakespeare is one of the best examples we have of a personality existing in words. He was a person of verbal rhythm, a master of words, and with words he created personalities. He certainly had a mind that used words to create, like the brain seems to use its chemicals and waves to people like us. His record of plays stirred up human nature, human drama, the innocent and the evil, the hilarious and the tragic, the living and the dead, the lonely and the loved, all of it, all of us. Can you see human personhood in Shakespeare's *Hamlet*, "What a piece of work is a man, how noble in reason, how infinite in faculties, in form and moving, how express and admirable, in action how like an angel, in apprehension how like a god? The beauty of the world, the paragon of animals—and yet, to me, what is this quintessence of dust?"

If the person is equally connectome and transmitter rhythms, computer and musician, and robot and poem, then it is true that expressions from science and the poet can equally reveal persons to us. We miss out on understanding our total selves if we are just exposed to our empirical selves. The scientist can see only part of the picture of the person if he reads only technical journals. Is it any wonder that our modern scientific culture, very shy on religion and poetry, has a model of man and his brain that is heavy with the computer metaphor and shy on even thinking about the person with a brain? For that reason the genius of Shakespeare is an admirable subject to see and to listen to on stage. At a time when English had a mere 6,000 commonly used words, Shakespeare showed us the insides and outsides of persons, as if he had crawled into their brains and reported the living emotions, wills, memories, and motivations of them all. He recorded them, not in PET scans and fMRI's, but in clothing, marriages, with swords, and with lips, all carried by iambic pentameter lines with their fleshy and bawdy words. And we remember his genius to this day.

We never had the chance to do either an autopsy or an fMRI scan on Shakespeare's brain. Sorry, but he died in 1616. A brain scan would be interesting, but perhaps no more an explanation for his genius than Einstein's brain at autopsy or the intelligent autistic, Colorado State professor, Temple Grandin's brain. There is more to William Shakespeare than meets your eye when reading his brief biography, and it is to be found in his words that carried images, emotions, and plots that open up and unfurl characters and ideas before your mind. His brain was different in ways we would not have been able to see. However, we do see something of William Shakespeare in all of his characters—Hamlet with thoughts of suicide, Juliet in love, Miranda in awe, Petruchio in his bravado, Beatrice in her longing … and others. Perhaps they all included pieces of his personality. How could we possibly squeeze him and them into his mere matter of brain?

Reductionism plus neurobiology are not the only way to explore a person like Shakespeare, or you and me for that matter. There is nothing in neuroscience that says you could squeeze the person and self-conscious mind of Shakespeare into mere neural networks, three pounds of genius brain, unequaled almost on the planet. Was Shakespeare's brain just a few more glial cells and longer fissures, and a bit more glutamate and NMDA receptors than mine or yours? There's the rub again. Shakespeare had very similar brain matter to all of ours, though very different circumstances were shaping its genetic and epigenetic form. His Wernicke's area of speech understanding, which is on the left side of most of our brains, was surrendered to a genius set of senses, a keen wit, a storyteller, an imagination from deep in the heart, and a personality lost in tragedy and triumph. Shakespeare was locked in his time, but he saw more broadly and universally into the ultimate and true nature of persons.

Concluding Thoughts

"I got rhythm" was a popular song from the hit musical *Girl Crazy,* in 1930 and a film in 1943 by the same name. The song was composed by George Gershwin, with lyrics by Ira Gershwin. *Girl Crazy* was about love and romance that can make any of us feel crazy. Rarely, if ever, do we feel as if love feelings are illusory or completely products of neurological action potentials. We are the dominant creatures on this planet by anyone's sensible definition of the term. Therefore, even though feelings are not always trustworthy, I will go with our experiences of feeling emotions rather than reduce love to an illusion or traveling sodium-potassium leaks on a cables of neurons. How can I bet against the genius and depth in the minds of people like Shakespeare or Einstein, or so many others who center their lives on the reality of personhood?

Mental life is a blend of thinking, feeling, and willing, and not just the action potentials that accompany these. The mind is related to but not merely brain areas that are knit together into complex networks. Yes, life is also seeing and hearing, but it is more your experiences of seeing and interpretations of what you see and hear. You do not see electromagnetic radiation, you see color, and colors you like and do not like. You do not even really see colors, you see colored things. Colors can disappear in about thirty seconds on a ganzfeld—a colored scene without any objects or borders on it. You do not see just women, you see the woman whom you love, and you see her in the rhythm of the feelings and thoughts of love, which are real. Life is subjective, and indeed it is related to the objective world of the brain with its neurons and neurochemistry. The subjective me is still related to my brain in ways we may never understand, but the important part of me is up there, moving around in my skull and body, invisible and undeniable. It is a shame if we are being told that everything that exists is in the objective realm—if the whole reason for the banning of the subjective conversation from the very beginning of the discussion on the brain is that being religious or romantic is just about an imaginary world and not the scientific real world. That kind of thinking, when you are living with your own feelings every day and with a three-pound wonder in your skull, is a short-sighted way to do the science of studying a person's brain.

Personhood is not just about the connections electrically and chemically between multiple brain areas. The secret to persons is not in their neurobiology, which is just the visible form of the person, like Shakespeare's characters are a form of him and a way for us to see his genius mind. Our biology of brain shows a wave form like fine handwriting, a rhythmic message form, and neither are we the pen and paper for the handwriting. But persons are the message of our active neural and chemical highways, a message we think and feel in our existence every day. The message is our very complicated selves.

Those who believe in God and the human soul have sometimes been accused of believing in a trick like, the man behind the curtain in *The Wizard of Oz*. We are being told today that modern science, instead of falling for the trick, has exposed those of religious persuasion, who have sold us a lie that there is a spiritual world beyond the physical. But is that really the case? If we examine our brains and our personal subjective experiences, then we have known the answer all along. It is no trick. We are persons and part of an immaterial reality. This is my guiding, top-down view point. I am not deceived. This reality is not an illusion. I am here, and I believe you are too. Every day we live with the fact of personal, subjective experience. That must be explained by those who think we are nothing but determined machines.

****

"Consciousness is the great poem of matter."
—Diane Ackerman (*An Alchemy of Mind*)

SOME BOOKS I THINK YOU WOULD LIKE

***Musicophilia*, by Oliver Sacks.** This is another important book by Oliver Sacks. His topic is not just music, but the important role music plays in the rhythms and functions of the brain. Music affects large areas of the brain and thus our emotions, memories, and movements. Music leaves its lasting imprints on our lives and aides in our understanding of how the brain may function.

***Neuroscience, Psychology, and Religion: Illusions, Delusions, and Realities about Human Nature*, by Malcolm Jeeves and Warren Brown.** These authors are both physiological psychologists with much work given to neuroscience. They give the reader a look at the issues between neuroscience, psychology, and religion. The book is readable, and it shows us areas that must be related if we are to truly understand our brains and ourselves.

CHAPTER 4

THE MOTHER AND CHILD BRAINS

Neuroscientist Rebecca Saxe recently had a picture taken of her and her two-month-old son, Percy, but it was not the kind of picture you frame and put on a side table in the living room. The picture was shot with an MRI's ghostly clarity, and it showed a mother holding her child. The MRI is not a quiet machine. So, with ear plugs in her ears and ear pads on baby's ears, she spent many moments over two days laying very close to her son in the machine, trying to keep him still. One day, when he fell asleep, she climbed into the tube of the MRI, held him very close, and the picture was shot. However, the MRI picture of mother and baby's brains was shaken out of the realm of scientific objectivity because mother Rebecca had leaned forward, to kiss her baby boy on the forehead. Overall Rebecca had spent hours in the scanner next to her child to get this picture, which beautifully blended her, an objective scientist with research on development in mind, with herself, a mother loving her baby. Art and science united, and humanness was joined with the brain.

A TITLE CAN SAY A LOT

The Ghost in the Machine

—Arthur Koestler

THE HARD PROBLEM: NEURAL PIXIE DUST OR GOD'S SPIRIT

"The brain can be weighed, measured, scanned, dissected, and studied. The mind that we conceive to be generated by the brain, however, remains a mystery. It has no mass, no volume, and no shape, and it cannot be measured in space and time. Yet it is as real as neurons, neurotransmitters, and synaptic junctions."

—Mario Beauregard

At a young age I learned that my eyes were windows for letting me see out of my head and into the world. My childhood self knew nothing of the problems with Cartesian dualism's theater of the mind. Of course, my theory did not tell me why I could see in my dreams or with my vivid imagination. As I grew older I learned from textbooks that the eyes only transformed information from the world of electromagnetic radiation into the electrical/chemical activity of neurons and synapses in the optical pathway and visual cortex of my brain. The picture of an action potential in neurons in my textbooks did not leave much of an image about the world out there of real things I knew I was seeing. The neuroscience book said that the visual cortex was responsible for sight. The text said it, and that settled it.

As I learned more I was further away from understanding how I could really see the beauty of fall leaves, hear the peacefulness of music, and taste a warm coffee with a friend, let alone marvel in the witty words on cats by a poet named T. S. Eliot. I did not see how electrical activity in neurons or groups of neurons could become cute kitties to me. To say that the frontal lobes were home to my thoughts and the limbic system was the source of my uncontrollable emotions was only labeling and not explaining how I was experiencing these things. Feelings from neurons did not make sense. I reasoned that I either had to close my text books or else go blind.

Our awareness of ourselves is not just in the objective biology of our brains, but in the subjective inner blend of thinking, feeling, and willing that accompanies us during our waking moments. This is not to say that my visual cortex does not provide objective information about what images are striking my retina at this moment. But it is the awareness of the information traveling from the visual cortex that I am asking about. I am aware of some things as just a background awareness, the sound of the microwave oven in the nearby kitchen, for example. Others I attend to and I am keenly aware of them and searching their details, such as the television across the room. I can also back up into myself and realize that I am aware of these things—self-consciousness is what we call that. Self-consciousness is my awareness of feeling things; the self-conscious state of being aware that I am aware of those things.

The Bodiless Brain

I recall years ago on Halloween watching with interest the headless horseman on his horse chasing after the hapless Ichabod Crane. This Halloween headless horseman seemed to see quite well without his head. I also remember some science-fiction horror movie about a human brain kept alive inside a jar of fluid charged with electric current, and the brain apparently could think with anger as it killed a man in the lab by thought alone. There was even research done during the time of the guillotine executions in France to see if the newly severed head still showed life and thought for a few moments after the blade fell. The evidence was ambiguous. Now, not in science fiction films, but in the pages of the May 2016 *Newsweek* special health issue, we can all read that within the year, medical science will be attempting the first human head transplant. There is already one volunteer for the operation. This is not some Frankenstein-like movie trailer where an evil, old, rich man gets a new body put together from spare parts, but it is current news. I assume that a large part of our consciousness travels with our heads, but we will soon find out if head transplants are possible.

Russian surgeon, Vladimir Demikhov, in 1959 transplanted a living dog's head and forepaws on to the body of another living dog. Two heads on one dog sounds terrible, even if it was medical research. We are used to kidney transplants, liver transplants, heart transplants, hand transplants, and even face transplants more recently. Why not head transplants? Actually, if it were called a body transplant, which is what it is, it would not sound so late-night science fiction. A healthy man has a crippling disease or a paralyzed body, but his head and brain are fine. Some other man dies of a wound only to the head, and his body is carefully severed at the neck and quickly frozen. Then, the head of our first man is carefully severed from his dying body and planted onto the new body, spinal column to spinal column, nerves painstakingly fused together. A little nerve-growth factor is applied periodically, with frequent physical therapy and some follow-up operations for the next year. At the end of that year our patient might be able to take his first steps. At first, complications make living very long a 50% chance. But after a decade or two, maybe the operation can become more commonplace. It took years for heart transplants to become a safer reality after the first successful human heart transplant in 1967.

To not think about human consciousness and the brain is foolish given what possibilities are emerging concerning human consciousness and the brain. It seems, though, that most of the time many neuroscientists have chosen to ignore the subjective experience of the conscious mind. This is because science cannot or will not deal with anything smacking of a soul, or will not deal with any problem that cannot be researched empirically or studied through sensory information. This problem of how we acquire experience or feeling from the functioning of mere matter has become such a hard problem that no scientist thinks it can even be approached with reasonable suggestions. Therefore, the hard problem, as it is labeled, is ignored, like a menacing sound in your car motor, which you hope will go away. The subject matter of neuroscience is me—my actions, motivations, emotions, thoughts, and mental illnesses. But since radically empirical science assumes that nothing of the subjective sort, like God, ghosts, demons, or personal experience, exists, any suggestions of that nature are just references to pixie dust mythologies. Many scientists will even ignore what is going on in their own heads, calling the subjective experience an illusion. A fine way to do science!

The Hard Problem

The label "hard problem" in neuroscience was popularized by Australian philosopher David Chalmers.[1] The whole topic had been discussed earlier in an article by Thomas Nagel.[2] The hard problem refers to explaining the phenomenon of our conscious experiences and why and how the objective physical activities of the brain's neural machinery should give rise to my subjective feelings. Such as the felling I get when holding my wife's hand as we walk, my seeing and experiencing the color on the leaves of trees and the beauty of the fall season, the meaning of declining ethical behavior in teenagers today, and the meaning of just about any experience that I consider important today. Consciousness can be defined as the elementary and more complex feelings that you have of actually seeing, hearing and feeling the real world, as well as reflection upon the beauty, mystery, awesomeness, and meaning of whatever is out there.

I am not a philosophical zombie or one of the new Japanese cashier robots with empty heads, nothing going on inside, *i.e.*, no feelings, no thoughts. I am conscious and I suspect that you are too. I suspect that a dog is conscious but not self-conscious, but I do not think a spider, with much less neural content in its head, is conscious. Actually, I can never know for sure what is going on in a spider's head, or a dog's, or for that matter your head. But I can know about my experiences, and personal experience is the most fundamental knowledge I can have. I am not only having these feelings, but I am aware that I am aware of them. I have self-awareness. There is something called the mirror test for self-awareness that babies pass some months after their first birthdays. Do they recognize themselves in a mirror. Yes, babies act like they are looking at themselves in a mirror after the first year. Gorillas and chimpanzees and dolphins and some other higher animals do. It is hard to tell exactly what the mirror test tests, but again, I can only really know what is going on in my mind.

Chalmers said there are two types of problems in neuroscience, the hard problem and the easy problems. The easy problems are by no means easy, but they are solvable questions about the brain's work in the production of learning, memory, hunger, sexuality, language, and much more. The hard problem is the problem of consciousness and, of course, that includes self-consciousness. What is the objective, brain-matter explanation for my subjective experiences, which are called "qualia?" (think "quality")? Rods and cones in our eyes are activated by light, and that activity is passed on to ganglion cells, and then to lateral geniculate cells, and then to cells in the visual cortex at the back of our brains. We can record every step of that process through the brain. But how does the action potential—that traveling leak of sodium and potassium ions moving down the optic nerve over 200 miles per hour and arriving at the visual cortex, and then on from there to other areas of the brain—create my actual seeing of red or black lines? Or why will neural firing of neurons in the auditory cortex create my actual hearing of a piano playing? I understand the vibrating eardrum and the wave motions of the basilar membrane and the firing rates of the auditory neurons, but how do those objective activities cause my conscious experience of the sound to happen? The sights and sounds are not just information for me concerning wavelengths of light, or air molecules bouncing off my eardrums. No, I really see and hear; I am feeling that information.

These neurological events do not give any clue as to why I am experiencing sights and sounds, let alone the pleasure of watching a baby laugh or listening to Dvorak's *New World Symphony*. If I know nothing else, I do know that I am feeling these things. How does a neuron or groups of neurons firing electrically create mental experience? That is the hard problem. In Thomas Nagel's words, "Consciousness is what makes the mind-body problem really intractable...without awareness the mind-body problem would be much less interesting. With consciousness it seems hopeless."[3]

If you are not willing to admit to the world of non-matter, or study the obvious subjectivity in your own experiences, or investigate the material and the immaterial in the brain as a reasonable possibility, then you have a hard problem indeed. That is where the materialistic, matter-alone side of neuroscience has left itself. If physics is the King of the Sciences, then neuroscience is certainly royalty, a Prince in the family of the sciences. Then, indeed, the prince has no clothes on. My experience of conscious feelings, or qualia, needs explaining. This is not just a theist's problem. Even hard-core, materialistic, reductionistic, atheistic, anti-Cartesian dualistic, no-ghosts-in-the-machine scientists will admit that this is a problem for any worldview, particularly a materialistic, all-is-matter point of view. How does a material brain in a material universe through material laws of cause and effect produce my non-material, subjective feelings? This is not just a hard problem; it is the hardest problem.

Attempts at Explanations

Why should we not just use the principle of Occam's Razor to settle the issue here, *i.e.*, when you have two competing explanations for a situation, the simplest one is most likely the best? In which case we should say that the simplest explanation for mind is hardly some hypothesized soul, spirit, or some other invisible stuff. Brain matter alone is all there is up in our heads and it will explain everything about human consciousness. But simple solutions are not always the correct solutions, and why would we expect them to be when we are dealing with admittedly the most complicated thing in the universe, the human brain? That the physical neuron is merely leaking out the mental life of a person is not a solution at all until some cause can be demonstrated other than correlation studies, *i.e.*, connecting neural activity with human reports of their experiences. To simply outlaw the hard problem by decree is dogmatism. Consciousness is the raw data, the most important data of all, to be explained, not explained away, and not explained as some side show of the brain's activities.

It is important to remember that correlations between active brain states and subjective experience do not answer the hard problem. Correlation does not indicate cause, although it does suggest the obviously close connection between the mind of experience and the brain's activities, whether neural, chemical, or subatomic. That close connection does not answer how an objective, physical, neural activity can turn into or produce an immaterial experience. The many correlation studies are helpful in seeing different areas of the brain's involvement in human behavior and experience, but they do not suggest what type of material cause there could possibly be to explain something not of this material, objective realm. And just ramping up the numbers of correlation studies does not change that fact! What this does mean is that our understanding of the human mind as merely a production of the brain is going to have to change. And, our understanding of the material universe, and science itself and its methods, may also have to change in order to even begin to investigate the hard problem.

Neuroscientists of importance searching for answers to the hard problem of consciousness include the late Nobel Prize winner Francis Crick, of the double helix fame, and his very capable colleague and researcher Christof Koch. They researched brain waves of 40 HZ (the gamma wave) accompanying conscious experience. They also described the mysterious claustrum in the brain that seems to act like an off-on switch to consciousness. Another well-respected thinker, astrophysicist Roger Penrose, has directed his search for subatomic ingredients and quantum gravity as key in the production of consciousness. But any studies to locate the neural correlates of consciousness (NCC) have not solved the hard problem, because the problem is not eliminated by finding areas of the brain to correlate with conscious experience. Deciding how those physical areas can possibly contribute to the origin of the non-physical conscious world is the need. At this point, even some hard-core, atheistic, materialistic scientists admit the hard problem is perhaps a problem never to be solved, and they prefer to move on to easier areas.

Then there are the emergent theories growing in popularity as a way to think about the hard problem. Interconnectivity of key areas of the brain as the source for our immaterial essence is thought to be important to conscious experience. Emergent theories suggest that the immaterial mind emerges out of the complex new networks or complex arrangements of neural matter in the human brain. The suggestion is that there are new qualities in matter that emerge out of combinations of matter, such as with water emerging from combining hydrogen and oxygen. Water is still matter just like hydrogen and oxygen are. The hard problem asks, how does subjective experience, which is not material, arise from the objective matter of complex brains? The problem is not solved by simply postulating that mind just emerges from combinations of neurons. That explanation is labeling, not explaining the problem.

Other answers to the hard problem include ignoring the whole debate as philosophical meandering in fantasy, pixie-dust land. But how can you say you are studying the person or the brain if you are ignoring the most fundamental experience of a person, that I am awake and I feel these things? I feel the touch of a person's hand. I hear the voice. I actually see the color red. I know why this joke is funny and I know what funny feels like to me. I can never know for sure what is your experience when you see red or laugh at a joke, or even that you are not an unconscious zombie, but I know that I am awake and have an individual experience of things in the world. Subjective experience is the fundamental data of my existence. This is undeniable, and it should be dealt with by the field of neuroscience.

Some thinkers, such as the philosopher Daniel Dennett, say that these conscious experiences are just illusions created by the material brain. They do not exist except as illusions. But this is only ignoring the problem or trying to explain it away. In any case I still see and feel my illusions. I still have the experience of the illusion, and how the brain produces the experience of the illusion is the question. Bernard Baars uses the similar terminology of illusion when describing the theater of consciousness, where there is no observer on the inside who thinks and feels. There is only a distributed set of neural cells that do get information from the outside, and those receiving cells are the theater of observation. But the hard problem remains. How do those systems of neurons produce the conscious experience I am having?

Of course, there is always the answer to the hard problem that we could label, "Be patient. We will figure out the hard problem just like we have figured out everything else." However, we must remember that the problem is the how of subjectivity, qualia, and experience. How does science touching only matter, deal with the non-matter of a mind? You cannot just ignore it with the prior assumption that all is matter. To run brute correlations of brain activity with conscious acting, willing, and feeling, does not show the cause of experience and never will. We all admit that the brain is closely involved with our experience, and thus correlations are to be expected. Just to run more correlation studies or to merely label something as a consciousness-producing correlate of experience does not explain to anyone how this happens in a material universe. Labeling is not explaining. What is at stake is the very real possibility that the universe is more than mere matter, and that our scientific methods need to take the subjective and the immaterial into account, if we are to ever understand the human brain and personhood. The hard problem is a demanding voice that must be listened to.

Michelangelo—A Person of Interest—God Confers Spirit on Adam

I do not know how exactly God's spirit created the immaterial essence of humankind, the beginning of conscious experience in all of us. But Genesis does tie God's spirit to Adam's body, and the two became one. I wonder if the beauty of God's creative ability is not admirably expressed in the arts, specifically in Michelangelo's *The Creation of Adam*, where Michelangelo may have been speculating on the human being as receiving God's spirit.

Michelangelo labored from 1508–1512 on the masterpiece called *The Creation of Adam*, the beautiful fresco on the ceiling of the Sistine Chapel in the Vatican. Medical doctors have reported seeing an anatomically accurate image of the human brain in the flowing cape behind the Creator God in this familiar painting.[4] People have wondered whether Michelangelo was making a statement about God conferring intellect on Adam in that moment, and not just life. After all, Adam is alive, his eyes are open, but he is lying listless with a limp wrist, barely reaching out to God. Michelangelo was a genius in his art, and his paintings were meant to affect us as students of art and theology. In the painting there is seen the outline of the frontal lobe of the brain, the cerebellum in the rear, the brain stem, and the pituitary gland in the back side of the legs of little naked cherubs facing away from your view.

This is interesting, but what earns Michelangelo a personal place here in my writing is not the brain on the chapel ceiling, but his brilliance as an artist. Yes, he could have been showing us a brain. He certainly learned about the human body through his dissections, but do not look past the genius of Michelangelo as you observe this painting. There seems to be more than brain at work here because he was exceptionally brilliant as an artist. He is a fine representative of art emerging from the human spirit and the need to reveal what is inside all of us. You have to go all the way back to the time of the Greeks to find equals to his great works.

Michelangelo was a superb artist, and his art showed us another way to see the world of theology. The 2014 film, "The Monuments Men," showed Hitler stealing for himself all the great art of Europe, trying to claim some dignity for himself and his Third Reich by picking up the art work after he had destroyed everything else. The most important art piece of the film was Michelangelo's statue of Mary, the Pieta, showing all the emotions of a mother, holding Jesus, her son, handed down to her from the cross of his death. Michelangelo's art revealed the whole person in us, and what we as persons needed and longed for in life. An MRI picture is hardly the art form of Michelangelo, but the mother and child at the beginning of this chapter carries the same emotional strength and love of a mother for her child. The art with its meaning is not just technique, but it is thought and feeling from within ourselves, and seems so far removed from the mere world of matter.

The unity of conscious experience involves another problem, which is related to the hard problem. That problem is called the binding problem: How and where do the parts and activities of your brain come together and function to make the unity of your conscious experience? Neuroscientists have worked on the binding problem because it seems possible to solve. You do not experience the world as just pieces of separate sensory events, but as a unified conscious experience flowing in time and across space. Information may come into the brain through separate channels, and there may be separate modules of the brain for different functions, but the question remains: Where and how do many sensations get put together to form our unified conscious experience?

Our conscious experiences also include the meaning of the percepts we somehow make from our brain's sensory input. Somewhere and somehow all of these individual percepts must be collected and painted seamlessly into one meaningful and deep experience. The binding problem relates to the problem of conscious experience—not its sensory pieces, but the whole world, the whole self, and your whole constructed version of a cute kitty in your arms.

Making Grandmothers out of Line-Detecting Cells

Let us begin at the beginning and see that this binding problem is a tag-a-long problem with the hard problem. The investigation of the visual system has been a good way to explore the brain since the visual neural pathway runs from the eye, to the center of the brain, and then on to the six layers of cortex in the back of the brain. David Hubel and Torsten Wiesel, eminent neuroscientists at Harvard, won the Nobel Prize in Physiology and Medicine in 1981 for their work on the visual system of cats and monkeys. They began by using single-cell recording techniques to record the electrical activity of individual visual cells, which exit the back of the eyeball and enter the brain. They were asking what visual stimulus in the eye made these neurons fire electrically. They would shine a tiny stimulus onto the retina of a cat or monkey's eye and then record from the ganglion cells getting that information from the retina. They found that a small dot of light or dark at the right location on the retina in the eye was the best stimulus for making these ganglion cells fire. Simply understood these visual cells were acting as dot detectors.

Eventually Hubel and Wiesel moved on in the visual pathway through the brain to the visual portion of the cerebral cortex in the back of the animal's brain. There they found that the optimal stimulus for vision, to make these visual cortical cells fire, were edges and lines of light or dark, in particular orientations. The visual system was not reporting the world out there to your brain like a camera would, taking the whole picture back, but it broke the visual world into tiny lines (visual analysis) and then later, presumably, it would have to reassemble these tiny lines (synthesis) into the world that you saw with your eyes. By the time Hubel and Wiesel were studying the cortical, line-detecting cells, they realized that they were seeing the synthesis of dots into lines, that is dot detectors combining to form line detectors. That was good news since they were getting to the combining of smaller features that eventually could bind together to make whole people and trees, and whatever you could see, and so on. With enough tiny lines in the right orientations and colors you could make any picture, and it was thought that must be how the brain does the binding. The brain simply binds individual percepts together to make the larger parts of a scene that we were viewing. This is what led to the search for the famous "grandmother" detecting brain cell. When you put enough line-detecting cells together you can make a cell that could respond specifically to your grandmother.

Grandmother Cells

Scientists in a 1959 paper, "What the Frog's Eye Tells the Frog's Brain," reported bug-detecting cells in frog vision that fired to small, dark, convex-shaped, moving objects—in other words, moving bugs. The different features of a bug were bound together and collected into one bug-detecting neuron. If human vision was at all similar, maybe we would have neural cells that would respond to shapes, and then combine those shapes into more complicated forms, until we had grandmother-detecting neurons. But alas, no one has discovered any grandmother-detecting neurons (the term coined in 1969 by Jerry Letvin), even though we can certainly see our grandmothers. The search was abandoned and the reason should be obvious. If you want a single neuron to respond to grandmother out there in your visual field, you would be asking a single neuron to do what we are not sure how a whole brain of nearly 86 billion neurons can do! And that is not all. How do you see grandmother when she is sitting down or when she is standing up? Do you need separate sitting grandmother-detecting neurons and other standing grandmother-detecting neurons? How about when she wears a red dress versus her yellow dress? And what if she is playing her violin while she stands near the grill in the back yard? You would need more neurons than there are stars in the universe just to have your visual system responding to your grandmother! The visual system has to create complex perceptions in some other way.

Jennifer Aniston Cells

It is interesting that the same impossible idea of grandmother-detecting neurons popped up in a new form. In 2005, the announcement came that scientists had discovered Jennifer Aniston-detecting neurons, and Halley Berry-detecting neurons in the brains of epileptic patients going through surgery.[5] Scientists did say that these neurons were discovered, not in the visual system, but near the memory areas of the hippocampus in the center of the brain. And the brain cells of these epileptic patients numbered in the tens of thousands of neurons that responded to the name, the image of the actress, the name of a movie she appeared in, and so on. These were not cells combining different features to make a unified picture, but larger concepts from large groups of cells used in memory and language.

The binding problem, therefore, has remained a problem, and we do not know how and where my individual sensory events from the world get put together into my holistic experience of reality. The parts of whole memories are even stored separately in the cortex of the brain. How are you able to see a cute kitty? You don't see the lines, you see a whole kitty. Lines are not cute. Our experiences are not just sensory, but they are meaningful, awe-inspiring, heroic, and all sorts of emotional and meaningful labels we can add. We do not find in the brain any one collecting area for the simultaneity and totality of our whole experience.

This discussion of grandmothers and kittens relates to the hard problem in that we do not see that the brain's activities relate to the connected world you consciously experience. Finding the neural correlates of consciousness or mapping the neural pathways of the brain will not reveal what/who is it that is combining the parts of the sensory world to produce your mental experiences. The cortex of the human brain has been mapped in terms of general areas, and no one area correlates with your whole experience. Whatever produces your conscious experiences is a creative factory of infinite storytelling. To say that all the cortical brain areas at the same time report their individual bits of the sensory world is leaving out the fact that our perception is not of a number of kitty parts, but of something entirely different from a set of parts. I do not even see the parts of the kitten, or the individual trees in the forest, or the individual parts of the automobile, or the arms and legs of my best friend. I see the whole scene in all of its context. The individual brain areas do not report the whole. Something must collect and work artistically and inventively to produce my resulting percept of reality. There is no scientist saying that in future years we will discover how the brain does the binding, because that still leaves the hard problem at the heart of the binding problem. And, thus, the hard problem seems here to stay.

To ignore the evidence that something highly unusual and of a different stuff is present in human nature seems inexcusable. And who can deny his own subjective experience? Everyone else in the room might be a philosophical zombie, but you know that you are not. We have to take your experience into account as we talk about the brain. The subjective is part of you and seemingly a fundamental part of the universe. Conscious experience is fundamental to who I am, what the universe is, and how we should be doing our science.

It is a self-imposed ignorance to assume that your knowledge of material reality exhausts all there is to know. To state from the very beginning, that matter is all there is, and all there ever will be, is itself grounded in assumptions about matter and knowledge. Such assumptions guide the choice of subject matter to study, the choice of methods, and the interpretations used to explain the research data, and that is circular reasoning in the use of evidence. The materialist says, "Matter is all there is, so I will use methods that only look for matter. And guess what, after many years of study, I have not found anything but matter." As we will see in later chapters, these assumptions about personhood also guide the choices we make concerning the applications of our science to pursue and what ethical guidelines are relevant to those choices.

Concluding Thoughts

The trend of some scientific researchers today is to call the reality of God a myth. That is scientism, or making a religion out of science. Facing off a generation ago against this trend was another head transplant story. In 1947, after witnessing the horrors of World War II, C.S. Lewis wrote *That Hideous Strength*, the third book of his science fiction trilogy. It's a book about science gone awry, and a scientific organization that was researching how to keep a human head alive without a body. The allure is obvious to those who wish to gain some sort of immortality. I would hope that our science is better motivated than the world of science gone awry in Lewis's fiction. Neuroscience, though, seems to be avoiding clear evidence of many types that human persons are not just material beings. If consciousness is the fork in the road of modern science and the scientific method, then we scientists have to be willing to add the view of consciousness in persons to our subject matter, and see how this fact might guide neuroscience research. Personhood needs to be a guide in neuroscientific theory and research.

There are different Christian views on the body, the soul, and the interaction between them. But there's also much agreement: We're persons created in God's image; we survive the death of our brains; and we'll be resurrected after death as bodies and brains, as full persons, living in eternity as physical, mental, personal, and relational beings. We do believe that we are created as persons in God's image, meaning that we are intended to relate in love with God and other human beings, just like God does. We are meant to be creative as God is, in our minds and in the work of our hands in His creation. We have agency. We are meant to be moral beings who see reality the way God does, and see with forward vision the way we are intended to live. We are also fallen beings with a brokenness that disturbs the body/soul unity, our physical lives, and our abilities and our relationships to God, self, and others.

The biblical description of the person says that we are embodied personal beings—spirit and matter—not a dualism necessarily, but the holism of Genesis 2:7, a created fusion of God's spirit and earth's dust. Genesis is not describing God putting matter and spirit together, as much as God transforming matter into persons (*nephesh*—the Hebrew word) with His spirit. Animals have *nephesh* also, but their *nephesh* is not said to be in the image of God or to be resurrected from the dead. We are not just immaterial beings, but we are a part of the physical world, embedded in bodies and brains forever, and, therefore, living through what bodies and brains can do for us in relationships and culture as a major part of our formation in morals, relationships, and the habits of physical and personal life.

Human beings are primarily relational in purpose, to God and to others. As self-conscious human beings we are capable of accomplishing this relational purpose with God's redeeming work on our behalf. John Donne said that "no man is an island," but in one sense we are, in our own heads, filled with private self-conscious experiences. In another sense we are not islands because, with the miracle of language, we can break out of our skulls and enter another human being's experiential world. We can also break out of our own self-centeredness, rise up from mere matter, and know the God of the universe. This is the biblical description of the importance of the immaterial mind and the self, all embodied in matter and brain, in human nature.

***

"Not everything that counts can be counted."
—William Bruce Cameron

SOME BOOKS I THINK YOU WOULD LIKE

***Consciousness: A Very Short Introduction*, by Susan Blackmore.** It is hard to beat the series of books called "A Very Short Introduction" on hundreds of subjects, published by Oxford University Press. This little volume by Susan Blackmore is no exception. She has the credentials in psychology, physiology, and consciousness studies to take this very complicated subject on the brain and its multiplicity of issues, and give you a readable, one-hour version, that is actually entertaining.

***The Conscious Mind*, by David Chalmers.** This philosopher and cognitive scientist seems to be the first person to introduce the language of the "hard problem" to neuroscientists, who have just ignored the importance of the hard problem. How our subjective experience arises out of an objective, biological world, however, is too important to ignore, since it suggests that we are beings existing in both material and nonmaterial reality. Chalmers gives a marvelous introduction to this problem of who we are as subjective beings, as well as objective bodies.

CHAPTER 5

THE INVISIBLE BRAIN

Multiple advances have been made in recent years in imaging the brain. One new technique for viewing the smaller details and circuitry of the brain has been used to turn mouse brains transparent. This is accomplished by removing the fatty lipids surrounding the neuron pathways. Neurons and synapses become more visible in that way. Light microscopes also become more useful in the research on the brain. Neuro images in this case better reveal the high-interest details of larger networks of neurons. The new method from Stanford University has appropriately been called *clarity*. Scientists aim to eventually make an entire human brain transparent. The technique will not replace other methods being used in the Connectome project, but it will supplement them in the studies on brain pathways and human brain diseases. See-through brains may be very useful to see neural connections, but looking through a see-through brain does not mean that we will have any clearer view of a philosophical question such as free will versus determinism. We may correlate neural firing with human decision-making, but we will not be able to see conscious willing with such methods. Freedom is never going to be visible floating around in invisible neurons. We could follow the famous Invisible Man by putting baby powder on the floor and seeing his footprints, because, after all, he still had feet! But why would we think that the hard questions about invisible souls and minds could possibly be answered by looking inside of invisible brains?

A TITLE CAN SAY A LOT

Who's in Charge?

—*Michael Gazanniga*

FREE WILL OR FREE WON'T: SOMEWHAT FREE AND SOMEWHAT NOT

"The more we discover scientifically about the brain the more clearly do we distinguish between the brain events and the mental phenomena and the more wonderful do the mental phenomena become."
—Sir John Eccles

Free will, when you attempt to study it, is a now-you-see-it, now-you-don't affair. Freedom is sometimes there and sometimes not, more like a Dr. Jekyll and Mr. Hyde locked inside of me, tugging and pulling against each other. That struggle in my brain is not to be viewed as a little red devil with horns and a little white angel with wings, both pushing me to do something. Neuroscience would generally say that it is more likely that one group of neurons is voting in greater numbers for "Do this!" than another group of neurons whose electrical/chemical impulses are voting, "Don't do that!" At times it seems to be my rational, prefrontal cortex versus my stormy, emotional limbic system. The decision to act or not to act will go to the strongest push from the strongest neural rugby players on the brain field. Discussing human will may seem to be a philosophical and religious topic beyond the purview of our brain studies, but that is hardly the case as some evidence will show. Free will does seem to actively make its presence known along the side of a complicated and constantly developing neural structure of learning and habit formation in our brains.

Maybe the lesson to continue to learn in our trek through the brain is to realize that we are incredibly complicated beings and brains. Nothing about the human brain is simple. Nothing about human nature, certainly not an issue like free will or determinism, is going to be easily resolved. With good habits, good parents, and exercising our wills when required, we do see success against our basest desires. We learn to exercise some control over our lives, and we often win in struggles against our bodies as we get up at five a.m. for prayer or jogging. Free will can give us the ability to win in the long run against competing thoughts and urges, and to act freely and humanely with our lives in the face of strong influences. Claims of determinism will have to be examined carefully given our inner experience of both freedom and controlling habits, and the constant tug-of-war battles inside of ourselves.

Questions about free will cannot be best answered with only human feelings, since I know that feelings can be in error. A high school freshman's feelings about the person she plans to marry someday are not that trustworthy. But neither can free will questions be answered with what is called promissory materialism, which says that free will is just an illusion created by the activity of your brain cells. "Give us a few decades, and it will all be explained," says the voice of determinism in neuroscience. "We have done so well with every other scientific question. Give us time. I promise. We will solve this one too." Hear the words of Francis Crick in his book, *The Astonishing Hypothesis,* "You, your joys and your sorrows, your memories and your ambitions, your sense of personal identity and free will, are in fact no more than the behavior of a vast assembly of nerve cells and their associated molecules."[1]

The Importance of Habits

One of the first things we notice about the human brain is that it is capable of taking care of most of our human activities on its own without any assistance from human thought or will, and it is a good thing that it can do so. Most of our learned behaviors become automatic over time, through repetition. I do not have to think about how to button my shirt or shave my beard. I just do it automatically. I can absent-mindedly drive home after work and arrive "miraculously" in my driveway. I sometimes wonder if I truly stopped for every red light! Indeed, I did so because my brain can handle most tasks of driving without me micromanaging the operation of a car. Years ago when my family moved to another home about eight blocks away, I thought I was driving to my new home, but I ended up in my former driveway several times. After the third time that happened my body got the message that we had moved, and I was automatically driving to my new home!

We have amazing brains that automatically and habitually run our bodies through the basics of thinking, feeling, and acting. Those learned habits serve the purpose of freeing up the human mind and will to think and make choices outside of and above the level of the mundane and the automatic in the world. To not have to spend time on the mental movements of tying my shoes, or playing chords on my guitar, frees me up to do the more important thinking, learning, and decision making of human life. My habits of behavior, when in place, will coordinate the movements of my hands as I tie my shoes, play the guitar, drive a manual shift car, and speak with scarcely a thought about the next word I use. I may seem like a robot at times until I have to learn a new motor skill. Or, many times I have to rise above my habits in order to change a skill, like an improper golf swing, or to stop a swear word about to emerge on my tongue, or to change a deep, rigid, erroneous way of thinking with racial bias.

Does this mean that we are determined and there is no free will? Of course not. It means that we human beings function in machine-like ways for the bulk of our behavior because it would be too time- and mind-consuming to have to think our way through all of our behaviors. We did think through the details of our actions the first time we were learning to shoot a basketball, to write with a pencil, or to use chop sticks when eating Chinese food. If, for example, 90% of our behavior could be classified as learned and habitual, then it is likely that the 10% left is used for free thinking and willing in more important matters. Habits of movement and thought can free up your mind's 10% to think with free will in a determined world.

The Backwards Bicycle

One of the more famous demonstrations of the strength of learned behaviors and the plasticity or changeableness of the human brain, is the backwards bicycle. Destin Sandlin, creator of the educational website called "Smarter Every Day," modified his bicycle so that the handle bars were geared backwards. Therefore, turning the handle bars to the right turned the front wheel to the left, and vice versa. Of course we expect that it would be easy to adjust to the new bike, but it took Destin a full eight months of daily practice to make the switch. It took Destin only twenty minutes to make the switch back to a normal bike. This backwards task is difficult because it involves changing a strongly ingrained habit of how we balance as we ride a bike. Destin's six-year-old son, who had already learned to ride a normal bike, took only two weeks to learn to ride a backwards bike. His learning to ride the backwards bike so quickly showed the amazing plasticity of the young brain to learn a different way to ride a bike.

The point is that most of what we do is a product of bodily reflexes and learned behaviors. You form new habits by repeating the learned response over and over again, thus strengthening the neural connections. With such a physical system in place, what you learn can function without your conscious interference. We tie shoes, eat with spoons, and ride our bicycles, all without thinking consciously about those activities. Most of our daily behaviors fall under that automatic-behavior label. Some of our habits are so strong that we can get caught when we get too close to some habitual behavior. This is called the capture effect of a habit. My mind was once deep in thought at home, and I happened to walk to my bedroom in the afternoon to pick up a book left there. But I did not look for the book; instead I "thoughtlessly" pulled my pajamas out of the second drawer of my dresser, all by habit. I was captured by the habit of heading back to the bedroom every night to get ready for bed. (I did not put my PJs on, by the way!) Someone could also get caught in anger situations where he starts shouting, and then he might start hitting instead of thinking about his behavior.

My purpose in this chapter is not to evaluate all the research on this topic of free will versus determinism, but to evaluate two classic experiments that have generated a lot of attention on the issue of free will. In this way we can see that the behaviors reported on are most likely coming from the determined, habitual aspects of human behavior. It is no surprise then that much of human behavior, and even some thoughts, when tested, will show themselves to be attached strongly to prior physical causes.

However, the real issue is whether there is human freedom present in our actions in spite of so much of our behavior coming from strong genetic, environmental, and learned pressures on thinking and acting.[2]

The Libet Experiment

Controversial research by the late Nobel Prize winning physiologist, Benjamin Libet, on unconscious aspects of human action, has been widely discussed on the topic of free will versus determinism. Libet, in 1983, used scalp electrodes on a subject's head to measure response times in the brain for a subject's decision to flex a wrist or a finger. In his studies, the measured brain activity occurred before the subject was conscious of deciding to flex a finger. These results were interpreted to show that the subject's free will did not exist. The decision to move the finger seemed to have already been made before the subject consciously decided to move the finger. A later study disagreed with this interpretation when neuroscientists Marcel Brass and Patrick Haggard found brain activity that showed subjects had the power to veto the actions they had begun. That response of subjects illustrating their free will was appropriately called "free won't."

Libet's subjects faced an oscilloscope timer while the electrical activity in their brains was being recorded with helmet-held electrodes. The subjects just moved their fingers or their wrists when they "freely" decided to, and noted on the timer (moving around a circle in about 2.5 seconds) the exact moment they decided to move their fingers. The electrical activity in motor cortex showed a spike in activity before the finger was moved. That is, about one fifth of a second before the finger was moved the subject's brain recorded the intent to move. In other words, a fraction of a second after you decide to move your finger, it moves. There is a brief delay for neural activity to begin finger moving. But (and this is important), about a third of a second prior to the decision to move the finger there was also recorded more brain activity involved in the initiation of the action. It was not expected that there would be a spike in activity before the subject actually timed the decision to move her finger. What the experiment apparently showed was that brain activity preceded the subject's conscious decision to move a finger by a third of a second. It was this research that led to the declaration that free will is just an illusion or an epiphenomenon caused by brain activity and not the cause of that brain activity. It looked like a scientist recording from your brain could tell that you were going to move your finger before you consciously made that decision to make the move.

Remember, there are a lot of unconscious reasons piling up for decisions I make willingly every day. These influences include past behaviors, learning from books and teachers, unrelated thoughts, my feelings of thirst and hunger, to just name a few. As I am making a decision, these unconscious motives rise to the surface to play a part and then sink below the surface again. Another study done recently with subjects who were in an fMRI scanner showed a good prediction of what the subjects would choose in deciding to either add or subtract two numbers for up to four seconds before they were aware of making their choices. The interpretation again was that free willing was nonexistent. This interpretation, which favors determinism, again ignores the possibility of many strong influences bubbling to the surface prior to a person making a decision. There are certainly other interpretations of these results as well as critiques of their methodologies. But first let me say that it seems clear that if free will exists in human beings, it is by no means a simple function. We have seen that much of human behavior may be determined by bodily reflexes or learned habits, but that fact does not take away the possibility of important free will actions and thoughts in many human behaviors and choices.

Interpreting the Libet Experiment

One critique of the Libet finger-flexion experiment is that the appearance of a spike in the brain wave, called the readiness potential, a half second before the action of moving a finger, does not mean that it is the cause of the action. The spike may simply mark the beginning of the forming of the intention to act, or the consideration of possible options. The readiness potential may be a sign that our free will is just ramping up to choose. Trying to decide when to choose to move the finger, while watching the timer to mentally record the time, is the complicated mindset of the subject in this experiment. This whole action of the subject is not simply a robotic movement of a finger. The subject's decision to note the time and move the finger seems to be a deliberated, mental action as well, probably in that case giving rise to the readiness potential prior to the official decision to move the finger.

Most behaviors where will is involved are not as simple as a single finger movement. The will for larger behaviors such as to go to India as a missionary or to do homework on time, are a function of many decisions over many years making you the person you are. Then, when the decision is made to take a course of action, it was probably already freely made much earlier, and the researcher could be seeing the choice in any situation or experiment as a determined act. Maybe this readiness potential is exactly what I should expect of simple, thoughtless decisions. Libet asked his subjects to let the urge to move appear on its own at any time without any pre-planning or concentration on when to act. Such urges that Libet was asking for are not easily thought of as examples of consciously caused events.

There should not be any complicated willing for moving a finger. Conscious decision-making may not matter at all in the milliseconds before decisions are made. If I had to go through some willing exercise every time I moved my fingers, I would not be able to do anything complicated, because I would be over-thinking every movement of my hand. With the finger, I just will to follow the experimenter's instructions, and that willful decision releases the well-rehearsed movements of my hand, which are then measured. Consciousness may not matter in simple finger or wrist movements as much as a willingness to open the gates and release the mechanical behaviors behind this simple act.

Our speaking rarely calls on our conscious attention to what we are doing. My spoken speech is perhaps a product of more freely deciding what to say being mechanically connected to words that I am used to using in such contexts. Deciding on choosing a watch to buy is a similar matter. And even then, the movement of fingers and wrist are already controlled by practiced, habitual circuitry for controlling my hand movements as I grab the watch and try it on. Even a choosing-to-add-or-subtract-two-numbers situation may only be showing an unconscious bias toward either choice. In that research situation the subject's choice was predicted at a 60% accuracy rate—only 10% better than the chance of picking one out of two choices.[3] Making choices are complicated actions and may involve a dozen or more mental and physical systems carrying out the choice, including some well-rehearsed habitual parts of the choices. Choosing simple behaviors and thoughts are important parts of this act of choosing. Typing on my computer keyboard is not a clear picture of my choosing particular words as I type. Consciously thinking through a course of action as I weigh options and consequences seems to be very different than flexing a finger.

It seems that studying this problem of free will versus determinism by working just bottom-up with tiny bits of behavior to understand larger behaviors is a problem here. Certainly, do not say you understand free will when you have studied only the bits and pieces of finger movements, because the bits and pieces certainly do not make up the whole of behavior. The whole behavior is admittedly far more complicated and spread out over much more time and over larger brain areas. Yes, most of us have less free will than we imagine, but that does not mean that human beings are determined. Our brains and conscious minds are the most complicated things in the universe, and therefore, after a look at a finger movement and a readiness potential, it is certainly premature to declare that human beings are determined.

Criminal Behavior

To understand the reasons behind crime and violent behavior requires that we again admit to the complexity of the human brain and personhood. We are not just determined machines, nor are we immune from the strong influences or temptations to violent or criminal directions. Human homicide has been around since Cain bashed Abel with a stone and on down to modern warfare's killing with its missiles from a distance. The oldest stone-age killing discovered was in an underground cave of almost 7,000 bones unearthed in northern Spain. The bones of 28 individuals who lived over 400,000 years ago have been studied and there was found the earliest evidence of a murder. A young man's skull was unearthed with two large holes over the left frontal lobe. CSI-type investigations have suggested that a weapon striking twice from different directions killed this man. Knowing the violence of people like we do, it is not hard to picture the scene in the cave. First, the two antagonists shout and threaten, perhaps over a woman, or meat, or leadership, and then one man strikes the other, swinging rapidly with some heavy weapon. Then, the struck man falls dead or dying to the ground. We can easily imagine violent men and such murders occurring frequently. Undoubtedly, future archeologists will dig up our own graveyards and find much more forensic evidence of our murderous cultures.

A more famous case in point of the brain and questionable responsibility for murder was that of Charles Whitman, a trained military sniper. In August of 1966 Whitman made his way to the top of the University of Texas Tower in Austin, Texas, and went on a shooting, killing spree. By the end of the day he had killed sixteen people and wounded thirty-one more before he was killed by the police. On the previous evening he had killed both his mother and his wife, and he left notes confessing to his crime, but no clear motives for the killings. An autopsy later showed that he had a tumor, called a glioblastoma, near his amygdala, an area of the brain often associated with anger and other social disturbances. But there was no agreement on whether the tumor was responsible for Whitman's actions.

The question of guilt hangs unanswered as we today have to deal with guilt in criminal court rooms where juries try to decide when human freedom leaves off and strong influences begin. The court's question is how much volitional control a person seems to have over his behavior when the murder occurred, and this becomes the information used in the sentencing phase of the trial. For Whitman, he would not have fared well given his very calm, rational behavior in the killing of his mother and wife, the leaving of the notes, his collecting his arsenal of weapons, and his choosing a high spot for his sniper shooting at those down below. But the truth is, we may never know how much influence verses free will occurs in any of our minds as we act. We do know, however, that relatively few murders are committed by individuals with brain tumors compared to the appalling number of murders committed in the United States each year. But genetics, alcohol, maternal neglect, physical abuse, gang pressures, violence in video games, access to guns, and cultural beliefs are suspected to play their own parts in human violence.

If our culture's deterministic, materialistic preferences continue to express the no-mind, no-free-will view, springing largely from the assumption of reductive materialism (you are reducible to mere matter), then the answer is already assumed, and it leads to no searching for responsibility in our behavior. Guilt or innocence is never the question. B. F. Skinner said it well with his book title, *Beyond Freedom and Dignity*, in 1971. There is no prize to be given to the gold medal winner and no guilt to be assigned to the criminal. We are beyond all that in our scientific knowledge.

All behavior is determined and we would be better off to counter-control human behavior ourselves rather than trying to decide how to change the behavior of apparently free people. This issue has surfaced in many books over the years, including Anthony Bugress's popular *A Clockwork Orange*, where the criminal Alex was thought to be just a biological (orange) machine (clockwork). The lawless young man, Alex, could be changed by scientific techniques, but if his personhood was real, the person of Alex would be lost in the therapeutic process. The idea of such determinism circulated widely with the startling neuroscience ideas of Jose Delgado in his book, *Physical Control of the Mind: Toward a Psycho civilized Society* (1971). Delgado was an excellent scientist whose work on brain control was popularized by his fighting a charging bull in a bull ring in Spain with no red cape, but only his wireless transmitter signaling to electrodes in the bull's brain. Once, in Delgado's lab where two women's brains were stimulated by similar electrodes, one woman said she had a desire to marry the therapist. The other expressed a great affection for her therapist. Physical attraction and marriage proposals are not by any means that simple, but the research was over-interpreted, spurred on by the assumption that human beings are simply material creatures led along by electrical potentials in their brains. Electrical control of criminal brains will never produce model citizens, just as Delgado's bull in the ring was never made into the peaceful, flower-sniffing Ferdinand the bull.

James Fallen is a neuroscientist at the University of California who has studied the brains of psychopaths. He discovered that his own family lineage was littered with very violent killers. Dr. Fallen knew that psychopaths had brain scans that showed very low activity in the orbital cortex, which functioned to put brakes on the amygdala in the limbic system. He then looked at a PET scan of his own brain and found the same low activity in the orbital cortex, just like all those killers. Fallen's genes also were tipping him to the side of violence like his criminal relatives. Fallen's question about himself was how did he avoid becoming violent like his relatives. Besides one's brain and genetic potential, it apparently takes abuse or violence in one's childhood to make a violent person criminal. Fallen's own family, however, had loving relationships with their children and extended family. Free will choices in this case were developed rather than beaten down.

Human Sexuality

Human sexuality needs to be thought of with the same rigor as the study of violence. It would be rare indeed to find any human behavior springing from small, single brain areas. Human sexual behavior is affected by a myriad of variables, and the bulk of these are not well understood in the infancy of our studies of the brain and human sexuality. To say someone is this way or that way in terms of their sexuality because of control by a specific brain area is not good science.

A highly publicized study in 1991 from Simon LeVay was interpreted well beyond the findings of LeVay. An overanxious audience seemed to be looking for evidence that men could be born gay. LeVay examined the brains of cadavers, 18 homosexual men, and compared these brains to the brains of 16 heterosexual men and six women. What he discovered in area INAH3 (the third interstitial nucleus of the anterior hypothalamus) was that gay men tended to have this tiny area of the hypothalamus resembling that same area in a woman's brain. With this in mind, it was called a feminized hypothalamus. This study was interpreted by many as proof that men could be born gay, and, therefore, they had no choice in their sexuality. Their homosexuality was natural and normal for them. The implication for churches was clear. Quit calling homosexuality a sin.

My purpose here is not to do an exhaustive exploration of homosexuality in this chapter's subject of free will versus determinism, but to try to make the main point on brain and behavior clear. I am not taking the time to critique LeVay's methodology, nor criticize him for the interpretations of his data that have come down to us from other interested researchers. This type of finding in the brain represents correlation studies between brain and behavior, and not causal relationships. We need to be very careful in our correlation studies that we not argue from just tiny spots or areas in the brain to complicated behaviors like human sexuality. We know from the history of phrenology how dangerous and, perhaps, how foolish that would be. Small areas of the hypothalamus are no longer seen as the centers of hunger because we now know how complicated the circuitry of something seemingly as simple as hunger can be. Why then should we replant the fields of phrenology with our far more complicated lives of sexual identity and behavior, or our complicated religious beliefs and behaviors? And yet, the gay spot with the feminized hypothalamus, and the God Spot in the temporal lobes, are still being used as answers for a variety of individuals working in science, who see only brain-caused behaviors in human beings.

On the religious side of this human sexuality question, Christians need to see the results as another series of experiments in many areas of brain and behavior that indicate how closely our behavior and sexual orientations might be tied to our brains as well as to our misbehaviors. It is not so difficult to look at the brains of some gay men in LeVay's feminized hypothalamus study and to decide that some people are born closer to the edge of homosexuality than others. We certainly would believe that about anger issues, extroversion/introversion personalities, chess playing ability, and other human behaviors and inclinations. That fact might suggest that we can see some of the human behavior we do not approve of with more compassion for the offending person, as well as hope that some of these behaviors can be changed in willing people. To believe that things are completely determined by brain mechanics, means little hope of change. To believe in total freedom means living with blame instead of working with the challenges of strong influences on behavior. Freedom of will in the midst of predisposing genetics and stiff environmental influences seem to be where human beings may find themselves today.

What is Human Freedom?

When describing the operation of free will in human beings, perhaps we should say that we are largely determined, but free when it counts. As with most things involving human beings, free will is not likely a simple attribute. It can be argued that "free won't," or something like that might explain the complexity of human willing. In that case, free will could function as a veto over the overwhelming number of impulses the brain might be bringing forth as choices. Free will then would act to veto every choice but one. That is why such a process is called "free won't." The left dorso-fronto-median cortex becomes active during such vetoes. Brass and Haggard, in an experiment similar to Libet's, gave subjects the choice to veto their initial decisions to press a button.[4] In this case a lot of activity is streaming through your brain about to affect your behavior, but your will acts to block everything except your choice of behavior. We do indeed experience the freedom to resist choices all the time. We choose to eat or not to eat sweets; to lose our tempers or maintain self-control. "Free won't" may indeed be a good look at an important form of human freedom.

One possible mechanism behind such free won't choices would be the concept of anti-memories being explored in neuroscience. It is well known that much of brain activity is operated by the balance between inhibitory and excitatory neural effects. When a system has an imbalance of excitatory transmissions, then activity occurs. When inhibition is brought into play, the activity of the system is slowed or shut down. In a new study from Oxford and the University College of London, it is this same principle that is being explored to explain memories.[5] The electrical activity making up a memory can be silenced when brought into contact with anti-memories made up of the opposite electrical patterns. The memory is not lost, just silenced. It is thought that this might be an operating principle present in much of brain-behavior activity, that is, altering the balance between inhibition and excitation in brain networks, can affect the particular activity expressed. In the case of free won't, the outcome would be that multiple impulses would be silenced by inhibitory replicas, while only the willed response would be expressed. Human will then would be suppressing all brain determinants but one, the choice you made.

Freedom seems to be a mixture of past decision-making in the middle of the complex world of genetics, culture, reinforcements, and the emotional push-pull present in human life. That description does not mean that everything we do is free. Actually, the opposite seems true for so many of our behaviors. But even those behaviors and attitudes can be intercepted and retrained by will over time. We seem to be free to rise up out of our material realm and use our small measure of freedom and mind to choose the good and the new and the different. Choosing to move a finger is not the same as a decision to go to college, to be a pastor, or to buy a house. Even a hemineglect patient's unawareness of his movements does not mean that free will does not exist at other times in that patient. Most of the time we are making habitual movements with no thought at all. Perhaps the biggest lesson here is that we have the opportunity to raise our wills against the pull of environment and genetics, and to begin to act with courage and determination when called upon. To be human is not to ignore biology, but to use freedom to rise above the more mechanical parts of ourselves.

John Polkinghorne—A Person of Interest

Free will versus determinism issues have produced some faith versus science debates that needlessly divide people of faith and people of science. One distinguished person of interest in faith and science issues is John Polkinghorne, who has had a distinguished career for many years as a mathematical physicist, and then for just as many years as an Anglican priest. For this reason, he has become a public spokesman for the dialogue between science and orthodox Christianity. Polkinghorne, from Cambridge University, produced dozens of papers and essays exploring the nature of matter in the universe, and was appointed a Fellow of the Royal Society in 1974. In 1982 he became an Anglican priest, and also served as president of Queen's College, Cambridge, from 1988 until 1996. He was knighted in 1997. He received the prestigious Templeton Prize in 2002. He is a humble man of superior intellect in dealing with the people and works of God's creation.

John Polkinghorne's beliefs allow him to study and live with a larger view of reality. He sees the world of matter and science as well as God and His universe as parts of the same truth that he is pursuing. One cannot read very much about Polkinghorne, the religious man with a Christian worldview, without picking up on his emphasis on bottom-up thinking from science. Most religious people have top-down theories and assumptions, and it is only by checking out these top-down ideas with bottom-up thinking from science and other experiences that one can have more understanding of what is truth in important matters. Polkinghorne stresses that learning is taking part in the experiences of others, and thus must be checked and tested with a variety of other experiments and observations. Science has demonstrated that it is good at this form of knowing. Top-down thinking in the church, for example, should be in a similar position. Our beliefs are based upon the experiences of others in the record of scripture and our teachers, and therefore should be subject to the same testing of ideas. And indeed we, when we are doing well, do read our commentaries, listen to the beliefs of others on the same subjects, check our facts and the ancient languages contributing to our beliefs.

However, the church in the past in its top-down thinking seems to Polkinghorne to have imposed its ideas on the world of science instead of learning from the evidences of scientific discoveries. Knowledge should be a two-way street between theories dealing with larger systems of the human person and culture, and the biological and chemical descriptions of cells and circuits in the brain. Polkinghorne would not agree with scientists who say that the standard ways of describing causation perfectly describe the ways that things and people interact. There may be higher levels of causation operating in the world of physical reality. He believes, therefore, that a theist, who is listening to the voices of science, can explain more than the reductive atheist, who often claims to have the corner on truth. To Polkinghorne, the physicist and priest, theism can make more sense out of the world of scientific data and higher levels of human experience than can atheism, which looks only to reducing everything to matter for the explanation of all things human.[6]

A good recommendation to Christians involved in science is to see reality from the top-down; not commanded down from the Pope or your pastor, but from a theistic worldview of God's creation, from our personal experiences, and from ways of knowing in addition to the sciences. At the same time we must do our work in science from the bottom-up, putting the pieces of reality together into a coherent whole. Wrong top-down theories must be corrected by good bottom-up data. And top-down viewpoints can give guidance and interpretation to bottom-up data. Top-down and bottom-up thinking, therefore, should never be separated. Thus, Polkinghorne avoided the narrow thinking of both Christian fundamentalists and reductive materialists in science.

In thinking about free will Polkinghorne believes that God's decision was that it was better to make free human beings who might sin, and not robots who were perfect, because only from human freedom could come real love. But our natures, including our brains, are linked to the physical world.[7] An examination of our physical natures should not hide the nature of God in us or the freedom to choose amidst all the strong influences around us. So often the bottom-up thinking assumptions of reductionism and materialism have led to a belief in determinism in human beings. Thus, it has been the assumptions of the reductive materialist and not the data of any experiment that have led to a belief in determinism.

Top-down, Bottom-up Attention

Because of the nature of brain function, top-down and bottom-up language also finds its way into the description of brain activity. The human brain is not just a system of neural networks with bottom-up stimulus input that are equal to the human person and behavior. Discussions of free will or thought-controlled action depend upon recognizing that such top-down connections exist. Researching bottom-up and stimulus-response activity makes one a good empirical scientist, but much of the brain does not operate in that way. Higher regions of the brain seem to have as much control over what is going on in the brain as do regions of lower-sensory input. What you see seems to depend more on memory, expectation, motivation and attention, for example, than on the sensory input pathways from retina to visual cortex. Human perceptual and motor functions seem to be at the receiving end of inputs from higher brain regions and not just the reverse. As has often been said, "We don't see things as they are, but as we are."

Top-down processing in the brain refers to activities that are initiated internally by human memory or expectations that affect the processing of lower-level brain activity. Some scientists, however, can take a bottom-up view of the task and say that they will study only the physiology of the brain, and then say that without any preconceived notions or beliefs, they will build up their knowledge of what the brain is all about by explaining higher levels of brain only by reference to lower levels of brain activity. However, this latter approach begins with top-down presuppositions, that is, assumptions about reality that are made prior to experimentation or examination of the subject at hand. Studying anything as complicated as the human brain needs to be studied with many ways of knowing that might help tip the assumptions behind research questions one way or another.

Every neuroscientist, including reductive materialists, does work top-down in the research enterprise in one way or another. Either assumptions are admitted into the lab and are, therefore, open to challenge, or one possesses hidden assumptions. Reductionism and materialism are themselves powerful assumptions that guide the research process in what subjects to investigate, what methods to use in the research, what interpretations to make of research data, and what applications to bring forth from the findings of research. Hidden assumptions like determinism and reductionism need to be examined like any other assumptions. Bottom-up thinking with a hidden assumption that all is matter is prejudging the case on human nature and not allowing neuroscientists to think in other directions and the general public to better understand the data we are finding on the brain. To gain a better understanding of ourselves from literature, history, religion, and the constant flow of conscious experience that we all possess would give a balance to the frequent misinterpretations of a reductive materialism present in neuroscience today.

Concluding Thoughts

We are not bodiless messages or spirits. We are embodied beings, spirit and matter, say our theologies. We are spirit-dust, spirit and matter fused together until death, and then fused again at the resurrection with new bodies. We are persons, individual messages being written, ink smeared across time's paper in a fallen world. When we are asleep or brain damaged, we are no less persons or less valuable as persons simply because the message has been garbled. My function or failure to function does not define who I am or my purpose or my significance.

We exist as an inseparable relationship between spirit and matter, thoughts and body, sparks and soups, electrical storms and chemical waves, rhythms and modulations of meaning and movement. We exist more as a concert, a dance, a novel, than as a coded information file. Just to make sure we all understand, broken hearts do really hurt, since pain is the result of stress hormones flooding our nervous systems, but unrequited love also hurts in more ways than physical. A brain scan actually shows that getting dumped in love has very similar effects on your brain as an addict's brain going through cocaine withdrawal.[8] An fMRI scan of you reciting poetry or Bible verses as you drown in pain is different from the normal scan of just the pains of stress on the body. Our embodied life means that we are like an ocean of salt water, salt dissolved in water, until the water dries up and only the salt is left, awaiting the life-giving water again and the resurrection of our bodies; for we are to be embodied persons forever.

****

> **"I here express my efforts to understand with deep humility a self, myself, as an experiencing being. I offer it in the hope that we human selves may discover a transforming faith in the meaning and significance of this wonderful adventure that each of us is given on this salubrious Earth of ours, each with our wonderful brain, which is ours to control and use for our memory and enjoyment and creativity and with love for other human selves."**
>
> **—Sir John Eccles (*How the Self Controls Its Brain*)**

SOME BOOKS I THINK YOU WOULD LIKE

***The Brain that Changes Itself: Stories of Personal Triumph From the Frontiers of Brain Science*, by Norman Doidge.** This is a book by psychiatrist and author Norman Doidge, who believes strongly in the brain's ability to heal itself. He shows the reader medical cases that tell exactly that. This is an interesting book about the brain's plasticity and its ability to change its own structure when needed.

***How the Self Controls Its Brain*, Sir John Eccles.** This book is by a Nobel-prize winner and an important forerunner in the area of brain and mind research. Though Eccles is often criticized because he is a dualist, he rejects any view that ignores the self, as he presents his case for you being the senior partner in the mind-body dualism.

CHAPTER 6

THE SPOT-LESS BRAIN

If you lost your glasses, you could buy another pair. Your wallet? That is more of a problem, but you could replace the contents. What if you lost a part of your brain? Such was the case of the lost vertical occipital fasciculus (VOF). Originally that spot was thought to be in the brain, and then it was lost. The VOF is a bundle of white matter fibers (myelinated neurons) near the back of the brain. Carl Wernicke, a famous name in the neuroscience of language, saw it in the brain and put it in his atlas in 1881. But any references to the VOF vanished from future brain atlases as neuroscience pushed toward today's supposedly more accurate versions. The VOF vanished, that is, until 2013. A Stanford graduate student at the time, Jason Yeatman, published a paper on his rediscovery of the VOF and its involvement in reading. Post-doc student, Kevin Weiner, went through the old atlases to find the path leading to the loss of this part of the brain. After some papers were published in the 1890s arguing against the existence of the VOF, the pictures, minus the VOF, in the early atlases became the standard of truth from then on. Those brain pictures began to affect what people would or would not see in the future dissections of the human brain. The VOF went missing in the brain in large part because it disappeared from the standard text books of the day.

A TITLE CAN SAY A LOT

Soul Dust

—*Nicholas Humphrey*

GOD SPOTS ON THE BRAIN: PUTTING GOD BACK WHERE HE BELONGS

> **"... religion cannot be reduced to a primary form of cognitive activity ... "religion" is more like "baseball"—a cultural and social phenomenon."**
> **—Malcolm Jeeves and Warren Brown**

The beautiful experience of listening to the soul-inspiring, Christmastime "Hallelujah Chorus" from Handel's *Messiah* arises from more than just an active neural network in your brain. We could say that each person's experience of this musical miracle is unique depending on the age and education of the listener. Trying to reduce the God-man Jesus Christ, to a single neural network said to explain everyone's belief in some god is an impossible task. To empirically define away the gods of Olympus may be safe, but to ignore everything beyond brain matter in the Christian religion's belief in God is a true head-in-the-sand approach to good knowledge gathering on religious experience. Surely, no one should say we are certain there is no God, given that we know so little about what can be known in our universe. Physicists, looking with mathematics and nuclear accelerators, say that 95% of the universe is dark matter and dark energy. Therefore, we must admit that we do not know very much about 95% of our universe, let alone the universe of conscious experience within our own heads. We still have a lot to examine about so much, and a little humility would be helpful in the learning process. Neuroscience needs to not throw out religious truth with the bath water of radical empiricism, but it needs to keep the God Spot conversation and the possibility of the existence of a personal, spiritual God within the bounds of humble soul searching.

Today we seem more enamored to explain God by searching out individual spots in the brain than by sticking with the well-supported view of the holistic, interconnected brain. Correlating the activity of brain networks to human behavior is difficult to do since we are dealing with a brain that clearly functions as a whole unit, and complicated human functions are not usually confined to single anatomical circuits. However, that has not stopped loud and proud reports on finding the source of religion and god in every human brain. What has been called worship of a god, for example, may only be what brain area lights up in a PET or fMRI scan when a person experiences something mystical or religious. However, we are forgetting that the whole brain is highly active when you are doing anything complicated, including anything religious.

The problem in much of neuroscience today is the tendency to assume something, *e.g.*, God, does not exist, and then proceed to use that assumption as truth when you find areas in the brain that you say show that god has been invented by man. The God Spot claim is that any brain activity when you are expressing religious beliefs must be the reason for the religious beliefs of people all over the planet and for all of human history. But there is no God Spot in the brain that can explain all the religious experiences on the planet, or that can determine the non-existence of God, unless with circular reasoning one begins with that very assumption—that no God exists. Circular reasoning, or beginning with assumptions that you wish to prove, is turning out to be a huge problem with neuroscience and mind studies. The very questions we are asking about the existence of the human soul and the afterlife, are the things that are assumed by some neuroscientists with the loud answer of, "There are no gods or spirits in our material universe."

However, keep in mind that the things in our experience of contemplating God, praying to God, or sensing the presence of God are certainly related to brain activity. We should expect brain activity to correlate with religious behavior. But why would we say that active brains are responsible for inventing the myth of God? It is possible that the reality of God and our responses to Him have led to changes in the brain just as much as violin practice or driving a taxi cab have been shown to change peoples' brains. You can simplify this whole question down to the chicken and the egg question. Which came first, chicken or egg? Did the human brain invent the God of the universe, or is God's presence responsible for the human brain that can know Him and relate to Him? To believe in God is not to deny the presence of activity in the brain that correlates with religious behavior, nor is it wrong to say these areas are useful and even necessary in our responses to that God.

The God Spot Cannot Be Just a Spot

No one area of the brain seems to do anything by itself, and we have every reason to think that the whole brain is involved in whatever we think or do. Whether brain activity on its own is the cause of the inventing of and worshipping of God all over the world is another matter (or should I say non matter?). What seems to be betting against the apostle Paul's conversion on the road to Damascus is the idea that God is just an evolutionary instinct confined within our brains, and Paul's declaration of faith, therefore, had to be something like an epileptic seizure. Brain activity in the hypothalamus may correlate with feelings of hunger, but hunger feelings and eating behavior are also correlated with the complexities of diet techniques, bulimia, metabolism, television advertising, the sight and smell of food, social activities, chewing behavior, loneliness, hormones, and many more things. When human beings are hungry, that hunger seems to relate to many other areas in the brain and body, in addition to what is happening in regions of the hypothalamus associated with eating behavior in animals. We have known that there is no specific hunger spot in the hypothalamus since the early 1970's with the closer scientific examination of the role of the VMH (ventral medial hypothalamus) in the classic fat-rat lesion studies.

So, too, far more than in the case of hunger, religious behavior and thoughts are related to cognitive, cultural, and emotional brain areas also known to function as part of clearly nonreligious behaviors. What is religion if not a complicated lifestyle, a longing for what is beyond self, hope in an eternal future, trusting in someone bigger than ourselves, finding some meaning and purpose to life, a relief from guilt feelings, a pursuit of sanctity and godliness in one's thoughts and behavior, prayer for the needs of others, and personal maturation as we age.

A language study recently showed fMRI scans of language activity to be occurring all over the brain.[1] The understanding of language happens not just in Wernicke's area (a well-established center for language) in the temporal lobe usually in the left hemisphere of the brain, but all across the entire brain. It was reported that hundreds of locations in the human brain's cortex activate to words with related meanings. This suggests that language comprehension is much more complicated than originally thought. The research team mapped thousands of tiny brain areas as subjects laying in an fMRI machine listened to stories. Individual words and the meaning of words showed up throughout the cortex and often in several places. There is no one spot for language that acts alone in the brain. Given the complexity of human language we should have guessed that all along.

Some say that religion is just an irrational side effect of the human brain trying hard to develop a completed package of explanations about the world around people. When we eventually understand our world, then these irrational religious belief systems will die out. The disappearing part has not happened in today's world when we understand so much more about our world scientifically and rationally. In fact, religious belief is held by many intelligent people, and across the world belief is increasing, not decreasing. Religion does not have a God Spot per se as much as it is just a part of our overall way of thinking about the evidences behind religious beliefs, about moral questions that confront us, and about the purpose and meaning of life, to name just a few. We do not doubt our abilities to deal with decision-making in many areas involving our beliefs and assumptions, so why would we single out religion and God as an evolutionary negative aftereffect, unless it is by circular reasoning that we have already decided what we believe about the existence of God and brain functions?

Active brain areas during religious thoughts and feelings are just about everywhere in the brain, which is what we would expect if religious activity uses the machinery of the brain. The important areas of correlation seem to be the frontal lobes, parietal lobes, and the temporal lobes including the emotional aspects of the limbic system. If a known language is involved, then language areas are active. If behaviors are involved, then motor areas are active. If a mental stupor or deep meditative state is involved, then we see activity in the temporal lobe and the brain stem. The frontal, temporal and parietal lobes are areas that handle many cognitive, social and emotional aspects of religious behavior and interact in many complex ways together. Some religious feelings are just very general, mystical-type feelings. But even these experiences can be built upon many brain areas that involve feelings of guilt, prayer, thinking, weariness, and hurt that may be leading a person to think about or to seek God.

After many years of research on this topic of the God Spot, we know that there is no god detector in the brain any more than there is a grandmother detector. The human brain, as we have seen, simply does not operate in that way. The human being is not a stimulus-response machine, but a thinking, feeling, acting being. To define us otherwise is by assumption and not by research. What the brain scans do show is that human beings, who have many and varied belief systems and activities, also show many interconnected brain areas correlated with religious belief, and those same areas are correlated with other non-religious human beliefs and behaviors. That idea should not be controversial by now.

There is good research on the topic of religious experience and the brain and I will review some of that here. My purpose is not to look at everything and solve some issue, but to suggest that good research can be done and is being done on this topic, and it does not have to be done beginning with the assumption that there is no God. Beginning with an openness on the existence-of-God question is a struggle for neuroscience. The neuroscience field needs to be fair and let the methods and ways of knowing of other well-established areas of knowledge contribute to the study of the human brain. Believing in the spiritual does not mean that the human brain plays no role in the whole of human experience. An eyes-wide-open approach to the study of the human brain may make the information we discover that much more accurate and valuable.

Temporal Lobe Epilepsy and Religious Experiences

The temporal lobes are the sides of the brain above the ears, including the inside of the same brain areas extending down into the limbic system in the center of the brain. Most epileptic seizures are of the convulsing-and-losing-consciousness types. There is, however, a problem, called partial seizure, which can cause a host of other symptoms including feelings of oneness, hallucinations, euphoria, out-of-body experiences, and the presence of god. Epileptics, who suffer seizures in the temporal lobes, occasionally report that they have intense religious experiences. Usually these are without convulsions. Many epileptics find meaning and enjoyment in these partial seizures. Good research on temporal lobe epilepsy and religious experiences was carried out by Dr. Vilayanur Ramachandran, at the University of California, San Diego, who presented epileptics with a number of images and measured their Galvanic Skin Response (GSR) or the changes in skin electrical conductance in the palms of their hands. This assessment is a measure of temporal lobe activity. Religious icons for these epileptics show more GSR than other images, so Dr. Ramachandran showed a link between the temporal lobes and religious experiences, and perhaps why these epileptics would have religious experiences during their seizures.

Temporal lobe epileptics have often reported mystical feelings and visual auras during seizures. Scientists used what was called a god Helmet that could produce a magnetic field and consequent electrical activity in these temporal lobe areas and the subjects would experience cosmic consciousness, religious feelings, or the presence of spirits. In an interesting test of what the god helmet was doing, the atheist Richard Dawkins, emeritus fellow at New College, Oxford, evolutionary biologist, and author of *The God Delusion*, donned the god helmet. Even though his temporal lobe was activated by the god helmet, Dawkins said that he experienced no religious feelings, only some changes in his breathing and feelings in his limbs. If God Spots created human beliefs in the spiritual realm, certainly Dawkins might have felt a little something, not like angelic voices singing "Amazing Grace," but maybe an aura or two.

It is well known you can produce religious experiences by activating the temporal lobes, but that does not prove that God is just an overactive neuron. In fact, people have all kinds of spiritual experiences from seizures and brain stimulation. When a certain brain area is not working in normal ways, unusual experiences can occur. This does not make such an area a god creator. I believe, rather, that God has chosen to communicate and relate to human beings most often in very ordinary ways, through the usual ways of the written word in language, through the lives of those who claim to have met Him, and in the reality of His becoming a human being in order that we might relate to the Divine God of the Universe.

Electrically Stimulating the Parietal Lobes

Andrew Newberg from the University of Pennsylvania injected a radioactive isotope into meditating Buddhist monks and found parietal lobe activity while these monks were in what has been called the "nirvana" meditative state. It seems more prudent to say that these areas of the parietal lobes are just performing a more generalized function of helping a person interpret and understand something that is not familiar or is a mystery, and thus religious beliefs can be affected. Just because this brain area is active during a religious feeling does not mean that there is no part to us that is capable of feeling and thinking deeply about the mysteries of life. Much of the wonder of the human brain shows that it is built to stretch far beyond the simple eating and breeding functions that we ascribe to its history. To call these religious functions, or artistic, scientific, and other higher human functions just side effects of the evolutionary urges seems to miss what a human being is.

This brings to mind the famous quote by Arthur C. Clarke, author of *2001: A Space Odyssey*, "Any sufficiently advanced technology is indistinguishable from magic."[2] Explanations of magic are probably happening daily now as people over fifty are having to deal with increasing levels of phone, computer, microchip, and other incredible technological advancements! Who knows what "magic" the next twenty-five years will produce in techy advancement and how our human brain will deal with it. Attacking science with accusations of demon technologies will probably become more common. But the opposite is also true today. Just because the mysterious and the spiritual is difficult to comprehend does not immediately have to relegate it to the magical world of pixie dust. Religious ideas can be well thought out, and there are intelligent reasons for belief. It should be clear that the skeptic's quote of "before you call something out of this world, just make sure it is not in this world," can work both ways.

Near Death and Out-of-Body Experiences

Near-death experiences (NDE) and out-of-body experiences (OBE) are also studied as a part of this religion-and-the-brain topic. At the point of death, people who have been "brought back" report feelings of floating, a strong sense of peace and beauty, seeing departed loved ones, a bright light, a long tunnel, and other religious feelings and experiences. Many of these people report changed lives because of their experiences. Natural explanations include that these experiences are the result of the dying brain's fight-or-flight response, moving into dream REM (rapid eye movement) visuals and consequent visual system activation, and a decreasing blood flow to the eyes. The fact that these experiences are much more profound and powerful than any other fight-or-flight responses or dream states that people usually have suggests more is involved than the natural explanations might be claiming. Any discussion of the existence of an integrated mind/brain and the actual existence of a spiritual world, would naturally be interested in a deeper explaining of these NDE and OEB events, rather than just explaining them away.

The brain slipping into a REM state of dream consciousness during a crisis in the brain does not seem to explain the great detail of these death or out-of-body states. Losing blood flow in the eyes may explain the long tunnel many people experience. However, it is difficult to say that these experiences are just dream states brought on by emotional responses, since these NDEs are much more real and profound than people's dream states. The person is "aware" of what is going on around him or her rather than just being "in" the dream, which is what most experience during their dreams. Even though brain changes accompany these crises in the brain, this does not mean that nothing spiritual is occurring during such states of near death. It is interesting that the people who are having these experiences generally report having a long-term, changed-life attitude afterward, seeing more meaning in life, and seeking relationships and goodness in life. Most of the time their perspective on their entire life has been changed.

It is interesting that when most of the brain activity has stopped at death, we can still have such vivid realistic experiences happening, almost as if there is a part to us that is not just bound to brain activity. In addition, there are those experiences that patients could not easily know about, *e.g.*, who was outside of the operating room, what someone said in another room, etc. But these are anecdotal reports and cannot usually be checked out empirically. The memories from these experiences, although they are episodic memories that are easily lost, are printed deeply in these patients' memories. Of course, strong emotional memories are "flash bulb" types and remain strong, but these NDEs show the same EEG activity of real memories and not just imagined scenes. A 2014 EEG study published in Frontiers in Human Neuroscience found that these outside-of-this-world memories gave off EEG patterns similar to real memories and significantly different from imagined events.[3]

Religious Experiences—Franciscan and Buddhist Meditation

It appears that in history God is usually making his appearance and voice present to us through our visual and auditory capabilities, and Christians believe that the same is true in the incarnation of Jesus and the inspiration of the Bible. In both of these God has made Himself a part of the world of material reality. This relates to the well-known Mario Beauregard and Vincent Paquette studies using fMRI scans to look into the brains and minds of Carmelite nuns who were asked to remember the most intense mystical experience they had ever experienced. These results were published in the journal *Neuroscience Letters*.[4] The scans of their brains revealed increased electrical activity and oxygen levels in at least a dozen areas of their brains. These brain areas were different than when the nuns were asked to remember their most intense emotional experience involving another person. Brain areas such as the orbital frontal cortex, which are associated with emotions in its connections with the limbic system, were affected.

Andrew Newberg, from the University of Pennsylvania, injected a radioactive isotope into Buddhist monks when they were in the meditative state. The parietal lobes were identified as active during their transcendent state. In their book Beauregard and D'aQuill commented on the results in this way:

> "... our experiments with Tibetan meditators and Franciscan nuns showed that the events they considered spiritual were, in fact, associated with observable neurological activity. In a reductionist sense, this could support the argument that religious experience is only imagined neurologically, that God is physically "all in your mind." But a full understanding of the way in which the brain and mind assemble and experience reality suggests a very different view."[5]

Andrew Newberg, physician and director (2006) of the Center for Spirituality and the Mind, has developed what he calls Neurotheology. His book on the subject is *Why We Believe What We Believe: Uncovering*

Our Biological Need for Meaning, Spirituality, and Truth. Coauthor of the book is Mark Robert Waldman. Together they have done brain-imaging studies on Franciscan nuns, Buddhists, and on tongues-speaking Pentecostal Christians. Newberg is not a member of any organized religion. He has used single-photon emission computed tomography, which measures blood flow in various parts of the brain, *i.e.*, more blood flow = more activity. He found no single area controlling religious experience. Our brains seem built to easily have religious experiences and pursue questions of meaning and existence, which often ends up looking for spiritual truth. Activity scans do not prove or disprove the reality of God. Science may never be able to answer that question. But we do know that religious experience exists historically and globally, and neuroscience may be an excellent way to study such complicated behavior. Hopefully, neuroscience should resist the urge to claim God Spots and deny that God exists based purely on the brain's allied involvement in such religious experiences.

Perhaps we need a new kind of science that can deal with the human brain and the many facets of human experience, including the spiritual. To just rule that human spiritual experiences are nonexistent by reason of empirical law is unreasonable. Such is prejudging the case about the existence of the spiritual and the role it plays in brain activity. We seem to need a new approach to neuroscience that would allow subjective epistemologies and approaches to human methods and data interpretation in order to give credence to the subjective side of brain activity. So much of the human mind is only approachable by subjective methods and top-down subject matter choices and interpretations. Consciousness and thinking is what the brain seems organized to do, and not just hunting down food and finding mates.

In every chapter I have brought up persons of interest, believing that all persons, not always geniuses, are key evidences for what the human brain and human nature are all about. This is not to dismiss the neuroscientific work correlating brain networks to religion, but to help guide our neuroscientific evidences into a better understanding of the thought and motives behind human behavior of all types. It is in this light that I want to bring up a few of the many examples of human faith that tend to be excluded today in the face of some neuroscientific proclamations on faith matters. I am choosing examples that involve, not just emotional feelings for religion, but complicated thought patterns behind the faith of human beings. The faith of these individuals are more than the firing of a God Spot. Faith involves rich amounts of thinking, feeling, and behaving, very often in organized religious settings.

Children of Faith—Persons of Interest

It is clear to those who study children that even the young can understand and feel drawn by the God they know. Faith in God can be seen in little children's words and drawings that have been gathered by Dr. Robert Coles, a child psychiatrist and professor at Harvard University.

His most well-known works seems to be his studies and books on the moral and spiritual sensibilities of children (*The Moral Life of Children*, and *The Spiritual Life of Children*). Children are not just indoctrinated by adults into religion, but they have many and varied ideas about their own faith and future in God.[6]

Dr. Diane Komp is a pediatric oncologist and researcher studying the faith of dying children and their families. This physician worked with children dying of cancer and got to observe their simple faith when faced with life's biggest loss. Her book *A Window to Heaven: When Children See Life In Death* deals with children's realistic approach to their illnesses and the difficulties and hopes they possess, in an area that adults find difficult to contemplate. The subtitle to *A Window To Heaven* shows an approach to dying that we would call true wisdom, except that it comes from our planet's smallest citizens.

Augustine of Hippo & Patrick of Ireland—Persons of Interest

These two very religious men were alike in that they were both sinners turned saints, and wrote books with similar titles about their turning to faith, *Confessions* by Augustine and *Confession* by Patrick. Augustine was the mentally tortured sinner in North Africa, and Patrick was a shivering shepherd slave in Ireland when they were called to faith by God. They both bared their inmost souls to God and to men. Both of their respective faiths were rational, emotional, life-changing, and became a lifetime of hard work and leadership. Both lived in the world and yet were not entirely of it. Augustine and Patrick, though they were near contemporaries, they almost certainly never met and they walked very different religious paths during their Christian lives. Yet, they left an intellectually and spiritually stronger Christian faith in Europe.

These two famous men of faith are interesting to study because of their connection to the fall of the Roman Empire and what we now know was a truncated Dark Ages, or a time after the collapse of the Roman Empire that was not so dark in faith and intellect as we have previously thought. With Augustine and Patrick we can see that the Christian religion and religious experience are a lot bigger thing than just feelings or mysticism, but their faith was a thinking and working faith. Their faith was thoughtful, and yet it was still a faith in an unseen spiritual realm. The faith of these men affected the world not through vocalizations and chants, but through the hard work and scholarship they inspired. An fMRI scan of their brains would reveal very little about their Christian belief in God. We can see the richness and complexity of their faith in their own words and the words of others written out for us to read and study. There is no doubt that faith was processed somewhere in Augustine and Patrick's complicated brains. But the faith we so obviously can see in all of its rational and emotional beauty is in their great minds and the continued work and growth of a church long after their deaths.

St. Augustine is considered one of the greatest foundational scholars of both Catholics and Protestants alike. Working from North Africa he penned his famous *Confessions* about AD 401, just about the time when Patrick became a slave in Ireland. Augustine converted to Christianity after a tortured life of sin, and he went on to renew the Christian faith during the chaotic, declining years of the Roman Empire. Augustine is looked back to so reverently because he was a mainstay of the Church as the Roman Empire and the Roman Church were collapsing under the onslaught of pagan invaders from the north. Augustine's book, *The City of God*, was important in that the city of man (Rome) was falling, but The City of God (God's church) was not going to fall.

Augustine is viewed as one of the most important Church Fathers of western Christianity. His theology and philosophical work preserved and seeded Christian thought for the next 1000 years down to Thomas Aquinas. Though he was Catholic, Protestants consider Augustine one of the theological forerunners of the Protestant Reformation because of his teachings on salvation and grace, so important to the Protestant churches. Augustine died in 430, just two years before Patrick arrived back in Ireland as Bishop to the Irish people. As contemporaries, the work of both would come together to help preserve the faith, culture, and learning of Europe, and helped to prevent the worst of the Dark Ages following the fall of Rome.

St. Patrick, living in the comfort of the Roman world in Britain, at age 16 was captured by Irish pirates and taken to Ireland as a slave. Patrick labored for six years as a shepherd slave, most of the time cold, hungry, and alone in the Irish hills with only the animals he cared for. In his isolation and suffering the young boy began to pray daily and became a strong man of faith in God. He escaped back to Britain after hearing a voice calling him home. At that time the largest Germanic invasion of the Roman Empire had begun and the Roman garrison in Britain was abandoned. He never again fit into the Roman comfort he had previously possessed. Patrick studied, became a cleric, and returned to Ireland in 432 as the Bishop of Ireland, two years after the death of Augustine. He had returned to a people who enslaved him, but a people he had learned to love. By his death in 461, Ireland was converted to Christianity, not by blood, war, and persecution, but by the faith of one man who loved the country of his slavery.

Rome fell to pagan invaders who began to burn the cities and the books of Europe's culture and accomplishments. At that time Ireland became the library of Europe, a safe haven to which to ship the scholarly books of not just the Bible and the Latin culture of Rome, but also books of the Greeks and of the Arab world. The Irish who previously were little warlike tribes almost overnight became peaceful scholars. As Thomas Cahill, author of *How the Irish Saved Civilization*, put it, "… as the Roman lands went from peace to chaos, the land of Ireland was rushing even more rapidly from chaos to peace."[7]

The Irish did not just store books, but they learned to read, copy, and translate them to other languages. Their monasteries became centers of learning in locations that became homes to hundreds and thousands who farmed and studied and worshipped in Christian peace. Today tourists to Ireland still visit Glendalough and Clonmacnoise and remember the scholars of Ireland. Within several hundred years the Irish were transporting their faith and learning back to Europe as they established monasteries all over the former Roman Empire. The former Irish war-like tribes were then not making war but were producing the beautiful art of the four Gospels in the Book of Kells, now on display at Trinity College Library in its famed "Long Room".

Faith and learning were preserved in Europe by the faith of both Augustine and Patrick. Their belief was not a carbon copy of the other, and yet their faith was in the same Lord of both heaven and earth. They had no constant auras or mystical feelings describing the entirety of their faith, but they showed in their differences a typical lifelong Christian mix of reason, emotion, and acts of obedience. Religion and worship of God is not simple, but always of the heart, mind, and behavior, made alive in the intricate cultures and diversity of the world's people.

Concluding Thoughts

Having brains that are active during religious experiences does not say anything about the existence of God or not. Faith in God does not exist simply because of an active God Spot in the brain. However, the biology of our complicated brains may strengthen our abilities to know our God; to rationally and emotionally seek and respond to Him. How else would we expect the eternal to reach out to the material man? We can know God, in part through the mystery of Jesus, God Himself living among us in history. We can know God with a mind/brain that can see and hear and respond to the presence and the pull of the spiritual realm of God, as well as the active presence of God in the Bible and the church today. Yes, it is okay to measure the brain just to see how it is active when we are having religious experiences. But to say that the brain activity is why we believe is again over-interpreting the data in favor of what is already believed about God, and our brain scans are playing the same phrenology game all over again.

The numerous scientific reports on locating the God Spot in the human brain have been used to say that religion evolved because primitive human beings could not explain the strange mystical feelings they were experiencing, and those primitives had to control their fears of the world of storms and large animals they were confronted with. Eventually, it is argued, these God Spot activities in the brain resulted in the billions of religious people on the globe, and the millions of churches, and multitudes of gods and sacrifices to legalistic practices to bring the faithful into line. Julian Jaynes, with his 1976 book *The Origin of Consciousness in the Breakdown of the Bicameral Mind*, even suggested that when figures like Moses, who during the period of time of the origin of languages began to hear voices in his head, he assumed that he was in the presence of invisible gods. Early man with his newly developed language centers began to hear himself talk and assumed in all this strangeness that the gods were talking to him.

Religion is a lot more like baseball in terms of human behavior than like surviving by hunting for food to cook and eat. Religion or spiritual experiences are not one basic, primary activity that can be located in one particular part of the brain or circuit of the brain. I am borrowing from an excellent example from Jeeves and Brown, and it makes the point well in our sports culture and bears repeating.[8] Our religious and spiritual experiences with God are more like baseball than like a God Spot. Baseball will not be traced down to a baseball spot in the brain. Baseball is a social phenomenon for those who play and attend. People can choose for a variety of reasons to like or dislike baseball. Baseball is cultural, tying many people together across broad stretches of time. Baseball involves players and coaches, and an audience, all a part of baseball. The game involves catchers and pitchers, and high-paid professionals as well as five-year-old T-ballers. Baseball is a topic of frequent conversation, traded baseball cards, and frequent purchases with favorite numbers on the back of jerseys. Baseball is a time of high emotion and a ritual time to share peanuts and hotdogs and beer, perhaps the only time you ever eat hotdogs. There are times for warmups before the game, a seventh-inning stretch, music before the game that is patriotic—the national anthem, the religious "God Bless America," and lots of country western tunes between batters. There are hours at home watching television with games 2000 miles away. Baseball is a social, personal, cognitive, and emotional pastime for millions of people. Baseball is not a religion, even though it has highly devoted fans.

Religion is like baseball in that many general, cognitive, emotional and social capabilities of the brain are called into play as we get involved in a life seeking after God. There is no baseball spot in the brain, and we find no God Spot in the brain either. Religious behavior does involve the motor cortex in the brain, a strip of cortex for moving arms and legs during worship. But is not the faith of many so much more than movement? Religious belief does involve feelings, but is not faith of the many so much more than electrical activity in our temporal lobes or limbic systems?

The Bible describes human beings as created to relate to God, in order to fill and fix an empty and broken nature. So often we have heard the quote that we have a God-shaped vacuum that can only be filled by God. This is light years from the God Spot theology. We are both material and spiritual beings, whether we are saints or sinners. In fact, we are all sinners. But we can stand and see the universe from our fallen place within it with our great capacity to know. If we were not conscious, then nothing is of any consequence or value, but because we are awake and can see and know about ordinary things and about God, then we can discover the meaning and value of ourselves as we relate to God. Suddenly there is a great drama to know and to participate in with our self-conscious minds. Rather than evidence of accident and lack of meaning, the fact that we are awake in a sleeping universe is the strong hint of our centrality to the meaning brought by God's existence and love for us. There may not be any God Spot, but according to many writers, we all have that God-shaped vacuum within and we are empty until it is filled. C. S. Lewis described this emptiness as joy in his spiritual autobiography, *Surprised by Joy*. To paraphrase Lewis, when we find that there is nothing in this world that can fully satisfy us, we can only come to the conclusion we were made for another world.

Blaise Pascal in 1670 published *Pensees,* which was his defense of the Christian faith. He said, "What else does this craving, and this helplessness, proclaim but that there was once in man a true happiness, of which all that now remains is the empty print and trace? This he tries in vain to fill with everything around him, seeking in things that are not there, the help he cannot find in those that are, though none can help, since this infinite abyss can be filled only with an infinite and immutable object; in other words by God himself."

***

"Perhaps religious belief causes the brain to change rather than the other way around."

—Andrew Newberg & Eugene D'Quill
(*Why God Won't Go Away*)

SOME BOOKS I THINK YOU WOULD LIKE

***Why God Won't Go Away*, by Andrew Newberg and Eugene D'Quill.** These two scientists used brain-imaging techniques to look into the brains of Franciscan nuns praying and Buddhists meditating. Their conclusions were that the human brain is changed by these techniques of intense religious experiences, and they are not merely subjective experiences, but the realities of God's presence being hardwired into the human brain.

***The Spiritual Brain: A Neuroscientist's Case for the Existence of the Soul*, by Mario Beauregard & Denyse O'Leary.** Neuroscientist Beauregard and journalist O'Leary put forth the case that religious experiences are not just subjective experiences created by the brain, but are actually created by the activity of God being active in the human brain.

CHAPTER 7

THE MINI BRAIN

Ohio State University scientists in 2015 grew for the first time tiny human brains from adult skin stem cells. The tiny brains do not have the architecture of an adult brain and are about the size of a five-week-old fetus, about the size of a pencil eraser. The purpose of the project was to make an actual human brain, though smaller, upon which research could be done to study human diseases such as Parkinson's or Alzheimer's, as well as a variety of drug therapies. Scientists could also study PTSD and traumatic brain injuries. These little baby brains contain 99% of the brain's cell types and genes, and even contain a spinal cord and a retina. To move past this point in the development of the brain, the lab would need a blood supply and a heart. These brains have no sensory input and presumably, therefore, no conscious experience. These are not science fiction brains floating around in a jar thinking, "Where am I and where's my body?" Developing these brains for research is important because scientists cannot ethically experiment on human brains except at death. Recent research with mini-brains shows the role the Zika virus plays on the fetal cortex. Three high school students played a key role in this research since they used a 3-D printer to produce a better bioreactor to spin the liquid and cells to form the mini brains. I guess youthful brain cells can oftentimes be better than the more weathered variety in most of our heads!

A TITLE CAN SAY A LOT

The Brain's Way of Healing

—*Norman Doidge*

PERSONS AT THE EDGES OF PERSONALITY: STILL THERE, JUST HIDDEN FROM VIEW

"A harp of a thousand strings is frequently out of tune."
—Sir Robert Boyle, 1665

Brains are not constructed with iron cables or steel I-beams like our modern skyscrapers. The brain's neurons and glial cells are not as solid as our new granite kitchen counters. A brain, sitting naked on your desk, would not last nearly as long out in the open as the forgotten peanut butter and jelly sandwich that you did not eat for lunch. Brains are loosely tied together and easily compromised when the connections between major cell masses are severed or malfunctioning. Some brain damage is chemical, when transmitter substances do not work or work too hard, or when the myelin covering many neural axons is slowly being destroyed. A stray bullet can be a major threat to your life as it causes damage in several parts of the brain at the same time along with nearby gushing blood vessels. An accidental fall off of a ladder while hanging Christmas tree decorations, striking your skull on the ground, can end your life in a few minutes. Whereas, if you lose a button on your shirt or spill coffee on your exercise pants, there is no danger to your life.

Though brains may be easily damaged, they are hard to destroy, because they are good at doing what brains do best, that is, doing things in parallel—doing the same thing many times, in many places, all at the same time. That is what is called strength in numbers. Visual information is processed over many different neural pathways running parallel to each other. Therefore, damage to one pathway of cells has little effect on the others. The human brain is good at repairing a variety of these brain damage situations. The main reason is because of what is known as the brain's plasticity, or the brain's amazing ability to change its own structure and function to make up for damage to or loss of brain cells and connections. Plasticity is available in part due to the brain's abundance of cells taking over damaged functions or even complex networks of learning. Through physical therapy the brain can recruit other nearby brain cells or areas to assist in recovering from brain loss—like a basketball coach who signals his backup center into the game when his starter sprains an ankle. This amazing plasticity gives us the ability to change and adapt as a result of experience or changing conditions. This plasticity, or being moldable, is especially active in the brains of children, but is still present to a lesser extent even in aging brains.

A recent example of research into tongue vision illustrates our brains' amazing plasticity. Blind persons can literally learn to see through their tongues! The brain of the blind person in this case is making use of cortical areas responsible for taste information and are appropriated for use in vision. The blind person wears a pair of glasses with a tiny camera mounted on them, pointing to the world out in front of the blind person. The camera picks up images and then sends those images to an electrical probe shaped like a flat tongue depressor held on the tongue. The images cause tiny pins to electrically stimulate the tongue in the shape of whatever is being viewed. The blind person may be looking at a letter of the alphabet or a plus sign, and that is the shape that will prick the tongue. What is interesting is that after a couple of weeks of training, the person really sees and experiences the shape! Areas of the brain responsible for touch on the tongue are suddenly recruited to create visual experiences. This is a compounded version of the hard problem! The point is that many cases of brain damage or disease do get better because of the amazing flexibility of the human brain. A similar example is haptic information, where the deaf person wears a shoulder vest that sends electrical signals picking up sounds to the skin of the deaf person.

Another example of plasticity is the case of the woman who was missing an entire brain part—no cerebellum. The cerebellum (which means little brain) in the back of the brain contains more than its fair share of neurons. Of the almost 86 billion neurons in the brain, the cerebellum would contain almost 50 billion of those. That fact makes you think it must do something important, and it does. It is responsible for musculature and motor coordination on both sides of the brain. And, some things about it help us see that it seems to confer rhythm on the brain, in walking, throwing, speaking, and literally everything we do. Life would be a dance if the cerebellum had its way in matters of function. But the point is that the cerebellum is extremely important to human activity.

Given the importance of the cerebellum, it was startling when it was reported from China that a 24-year-old woman was discovered who was born without a cerebellum, and had lived a fairly normal life. She did walk a little unsteadily and has slurred or halting speech. The condition is known as cerebellar agenesis and is extremely rare. Usually this condition has been found in infants with mental impairment, although only discovered through an autopsy. A CAT scan of this woman's brain showed nothing but cerebral spinal fluid where the cerebellum was supposed to be. How was this woman living without this important part of her brain? We do not know. This again supports the incredible importance of plasticity of the brain. And, when we see any brain-damaged person, there is always the thought that through physical or occupational therapy, or on its own, the person can recover some function.

The Acquired Savant Syndrome

Problems in the human brain do not always result in what we might consider something negative to the person. Brains for whatever reason might also ramp up the level of our brain's potential as in the case of acquired savant syndrome. Such cases illustrate that brain damage might be a key to finding the genius present, perhaps in all of us. Derek Amato, at age forty, damaged his brain diving into a shallow swimming pool. He essentially came out of the water with a severe concussion and was suddenly a musical genius. How this happened no one knows, except that it may be that the shock to the brain itself may have rewired the chemistry of the brain to see and work with music in different ways. Doctors at the Mayo Clinic said that he may be one of a very few individuals in the world to have genius level skills released by a concussion. The scientific explanations for Derek's sudden savant musical ability suggest that as Derek's left hemisphere of his brain was damaged, the right, artistic, musical side to his brain was released from the "tyranny of the left hemisphere" and now ran free to generate thoughts with music.

Another mild concussion producing an art savant took place in the brain of Jason Padgett, who was kicked in the head by muggers. When Jason recovered he began to see fractals everywhere. A fractal is a geometric curve each part of which has the same pattern as the whole. (You have to see one to really get the idea.) He became a math and art genius almost overnight. And then there was the Italian immigrant, Franco Magnani, who in his thirties came down with a severe fever which resulted in seizures and vivid dreams of his childhood home in Italy. Franco, not a painter, began to paint almost with an obsession, very detailed thirty-year-old memories of his home town down to the smallest details. In another similar case reported in a *N.Y. Times* article, Lawrence Osborne had his own brain stimulated by a Transcranial Magnetic Stimulator. Immediately his own drawings of a dog improved dramatically as the exposure to electrical stimulation of his brain increased. When the current was diminished, the art ability also decreased. Somehow, damage or changes to the brain can unleash buried abilities. Derek's and similar cases do show us that we do not know very much about the brain and our minds, except that personality problems may be better thought of as altered pieces of artwork rather than as busted computer programs.

Problem Brain—Problem Personality

Our brains are not perfect, and problems in brain function can produce abnormalities in one's personality. Tumors or drugs become the usual suspects when strange behaviors occur in otherwise sane individuals. A young man in the UK is reported to be living in a world of persistent *déjà vu*, where he finds himself constantly feeling like he is reliving events from his past in his own personal movie. These episodes began when he began taking LSD during his first year of college. We all have experienced *déjà vu* for brief moments, but fortunately not over long stretches of time. Brain scans do not reveal anything out of the ordinary about this man's brain, and yet his personality is becoming filled with fear and anxiety. We do not know why *déjà vu* experiences occur, but it seems reasonable to suggest that these feelings might help us when we are trying to recognize a familiar person or recall a previously visited location. We have the natural memory experience of having been there before or having seen that person before. Upsetting the function of a memory system in the brain obviously produces some very uncomfortable abnormalities in this man's personality. The point is, when our brain is hurt, then our experiences, even the conscious experience of ourselves, can be radically altered. It seems that if we become unable to process information in familiar ways, then we become personally damaged.

Changes in the human personality in an aging brain is a common example of the close connection between the self that we see and brain matter. We are all spread out over time, and vulnerable to damage anywhere along the way. I remember my mother when she grew older. She was not always the same person I had known all my life. Her personality grew dimmer, or thinner—I do not know how to say it. But she was still a person of value and my mother. We do not look at aging, medal-winning Olympic athletes and expect them years later to do gymnastic routines with the same skill as in their youth, or to do the 100-meter dash with the same speed. We are persons over time and with very important but vulnerable brains. Nearer to the end of her life my mother looked more and more as if she did not remember who I was or how many children she had. One time she even pointed to a picture of my dad and swore that was me. After a minute or two I stopped trying to convince her otherwise and changed the subject. She was still my mother, but not full of all her former personal abilities and memories.

I remember the pictures of Ronald Reagan as president, full of personality and authority. And then there were those pictures of him when he had Alzheimer's, standing meekly and blank faced beside his wife Nancy Reagan. He was gone if you mean all the personality attributes he had possessed, but we would not say he was not a valuable person deserving of life and respect and care. The person of the Ronald Reagan was still there, but the way people were used to interacting with his personality was now gone. So also it was with my mother. Like Ronald Reagan, my mom was missing some personality attributes, but she was still a person nonetheless, largely hidden from view, and it was up to me to find new ways to interact with her.

The same idea of still being a person but hidden from view can be used for persons in coma or vegetative states, or other physical, mental, and emotional problems. We are all vulnerable to personal damage as we move in aging toward death, the ultimate loss of expressing our personalities here on earth. Human beings are created in God's image as persons, and it is from that fact that they draw their value and purpose. Their personality attributes, or lack of the same, do not define their nature as persons. Persons have a declared value rooted in their Creator, who declares His love for them, young or old, conscious or unconscious, sane or mentally impaired.

Looking for Personhood in Damaged Brains

We normally define personhood by more than brain activity. We speak of the attributes of persons when they think, communicate, create, make moral decisions, relate to others, and are self-aware. An even more important general attribute of persons is in terms of relationship; the ability to relate to other persons, and to be in relationships with others. This includes the ability to respond to the personhood of another human being, and to show moral responsibility to others, and to God, since He is also a person.

These attributes are certainly good for as far as they go, but often not helpful for deciding what value has a brain-damaged person, or a person in a coma, or an advanced case of Alzheimer's disease. To consider that personhood is nothing but a product of biological matter and human activities leaves us open to just making an arbitrary decision for who is still a person and who is not. It is not true that if the person is in a deep coma, or is not talking, not aware of self, not thinking, then the person does not exist and can be extinguished or used for research purposes. Severely damaged persons are still present as persons even though their abilities to communicate and relate are absent. Brain damage does not destroy personhood. It just pushes personal expression out to the far edges of personality, largely lost from view. Personhood needs to be expressed with a healthy brain, but it is not lost nor is its value in the damaged brain. If we are deep, complicated musical instruments, then we can be deeply hurt in complicated ways that conceal our musical songs.

Our value and significance as persons is not in the abilities and attributes that we have and show, but in the value we have been given by the God who loves us. The female who is beautiful may feel more value than the one who is not beautiful. The tall athlete is treated with more value than the less-athletic student. But when all these people are someday over one hundred years old, none of them will feel any value because of appearances. Their value at age fifteen or one hundred years old needs to be in what is called declared value, something unchanging when someone important says you are valuable to them. If the mother of an average-looking girl says to her daughter, "I think you are pretty," the daughter would probably say, "Thanks, mom, but I want boys at school to think I am pretty." The girl's value is in the one who declares her worthwhile. And what we learn from scripture is that the God of the universe values us. And that fact will not change when we are one hundred years old, or even for eternity. Our personalities and worth are outside of what we can think and do. We are persons of value and related to the God of the universe, even if we have damaged brains and personalities.

Brain activity is not the person. And neither is a dead body a person. Of course it is important medically to know when a person's brain is dead so that we might arrange organ donations or arrive at decisions on when to pull the plug on physical life. We also need neuroscience research to learn how and when to improve our brains by adding neurons or electronic chips to our brains to improve our functions. Scientific data can help inform our moral thinking in these matters of brains and personal life. These are person improvement questions and not usually "who is the person" questions. Improving persons is much more than improving my sensing and thinking functions. If we were to emphasize improving human function, the top of the list would be to stress relationship abilities, which rely on self-awareness, communication, rationality, and free will.

Persons Out on the Edges of Personality

My purpose for the rest of the chapter is not to explore the causes or cures of mental and emotional problems, but to consider a few examples of persons whose brains are not functioning properly. These people are not just a list of symptoms, where each person represents his or her own pages in the Diagnostic and Statistical Manual of Mental Disorders (DSM). Some have memory problems. Some have communication difficulties. Some look with empty eyes. Some are in bed and eyes are closed. Some eat only what's on the left side of their plates and dress only the left side of their bodies. What is wrong with these people? They have damaged brains and those damages cripple some of their abilities to perform the way they used to or the way we expect persons to. What we do not need to do is to think that they are not fully human persons in these or other cases. Their personhood in each case is not gone, but just hidden from view. They are unable to show their whole selves through their damaged brains, being out there on the edge as they are. In this life of personal troubles and a relentless aging toward death, we are all in some way showing some less shiny side of our personalities. It may not feel like it, but we are all living somewhere out toward the edge.

Maybe the question to ask is not when personhood is not good any more, but how should we be dealing with the less visible or broken personalities around us. We do not see infants as whole persons, but we care for them. We even talk to our little "munchkins" in what is called "motherese," with simple grammar and exaggerated melody. It sounds ridiculous when moms in the check-out line of the grocery store say, "Oh Poopsie, look at the little doggie!" That caring mode of speech is good for the relationship of both the mother and the child. We can relate to all brain-damaged people, no matter what their condition, even though they may not be able to relate to us. Is that not the way we approach the poor and the sick? We should not ignore them simply because they are weak as persons, but we should take their condition as the opportunity to act as persons toward them out on the edges of personality.

Temple Grandin—A Person of Interest

Autism or the autism spectrum is a single name, but there are many individual differences on the spectrum. Several disabilities seem present for most autistics: a profound lack of social skills, poor communication, repetitive behaviors, and many times sensory oversensitivity. While it is a spectrum and individuals vary a great deal in these four, all individuals with the disorder have social intuition problems. The severe autistic is not very good at sensing the mind and thoughts of other human beings. The autistic cannot put himself or herself in the mind of another individual whom he may be observing. She cannot easily guess what another individual is thinking. The autistic has little theory of mind. The autistic may lack a class of neurons called Spindle neurons, or often called Von Economo neurons (after their discoverer), which are found largely in frontoinsular and anterior cingulate cortex and seem to aid in social learning. Autism may work to disrupt this learning. In addition, mirror neurons, in premotor cortex, which seem to be important in perceiving and understanding the actions of others, may not be functioning properly in the autistic person. If this paragraph sounds like the DSM-V, let me move on to the person beyond the symptoms.

One of the more famous people with autism is the highly intelligent autistic, Dr. Temple Grandin, author of *Thinking in Pictures* in 1995, and more recently *The Autistic Brain* (2013). Until her recent retirement she was a professor at Colorado State University in the Animal Science Department. Temple is not just an autistic, but an autistic savant, which means she has an extraordinary mind with her photographic memory and excellent spatial skills. Temple has some difficulties processing words, but the fact that she thinks in pictures, helps her make up for this language difficulty. It is this type of thinking which she believes helps her understand what animals perceive, and thus she has become an expert in animal care. In her professional career she developed animal handling systems for reducing animal suffering.

Temple Grandin's thinking is so different from the students with whom she works that she has had to memorize hundreds of social situations that help her anticipate what students are thinking, and how to best deal with social situations. In fact, she herself thinks she is more like Data, a character from the "Star Trek" television series, who is an intelligent android without human emotions. Since it is the emotions more than reason that help develop personhood, the autistic is less able to enter the full drama of what a person can exhibit.

Temple Grandin's brain has been scanned dozens of times with everything from the ever useful fMRI to a diffusion tensor imaging (DTI) scanner. Her scans show a larger ventricle in the left hemisphere, which means a larger Cerebral Spinal Fluid (CSF) filled space in the brain. When the brain has lost brain matter, the CSF fills up the space. This suggests that her left hemisphere has abnormalities resulting in brain tissue loss. This is a possible reason for her difficulties communicating in words and especially with human beings. She thinks in pictures and that may help her in her work with animals. Her brain also shows a larger amygdala than normal in the limbic system's emotional-motivational circuitry. This larger amygdala may explain her affective emotional states, in particular her lifelong anxiety. Temple's fusiform gyrus, in the inferior medial temporal cortex, is smaller than normal. This part of temporal cortex is the region associated with facial recognition, a social skill often lacking in autistics.

So, what is it like to be Temple Grandin on the inside? We cannot know, but we can guess. And she can tell us. And it is certainly true that what is in her autistic mind is definitely not the same as in other autistic minds. She, like most autistic persons, has sound sensitivities. Because of sensory processing difficulties she is perhaps not only not receiving sense information the way you and I are, but her brain is also interpreting sound differently. In which case Temple Grandin can be said to be living in an alternate reality from our own. It is hard for an autistic person like Temple to live in our reality that does not seem to tolerate her sensory processing and it is screaming at her in many different ways. To evaluate a temper tantrum in an autistic child from our point of view is to not do justice to the experience of an autistic who may be exposed to sensory overload.

Temple does not think that most professionals know what an autistic world of experience is like inside the mind. She lacks a theory of mind, meaning that she does not have to ability to imagine looking at the world from someone else's point of view and to respond appropriately. Like most people I am not autistic, but we who have a theory of mind fail to consider what things are like from the autistic point of view. People it seems have different brains for learning and processing information. A racial bias is a way of processing. So is being a geologist at the Grand Canyon, or a kid in a candy shop. But the autistic is an even more unusual person, one almost not built for this world, or not social enough to get along in this world.

Another autistic person of interest is Daniel Tammet, the British autistic savant and the author of *I Was Born on a Blue Day*. Daniel has difficulty interacting with and communicating with others. Daniel is the man who memorized the symbol Pi out to 22,514 digits. (I only know four—3.147.) How did he do it? Daniel imagined a color or a shape for every number. How many colors and shapes can there be? He obviously has a different brain and genius not available to most of us. Daniel said though that his true ability was in languages, and in response to a challenge to learn a language, he learned a difficult language, Icelandic Nordic, in a week. It is spoken by only six million people in the world, and now six million and one. Daniel walked around Iceland with a guide, and though he was ill at the time, he was interviewed on Iceland television at the end of the week. The interview of course was completely in Icelandic Nordic. All that language ability but less ability to use language to socialize with the rest of us. And thus he remains out there on the edges of personality.

These two examples show how easy it is to be human and yet to be very different. In fact all of us are very different from each other. And that is what is important here. Those with autism may seem to be asocial and love being alone, but with most it is a problem of not fully understanding the social world around them. The truth is that they are not as equipped as you are to join in our party. They do not know how to make friends and have relationships. The trouble with being autistic is that we are not living in an autistic world but in a world run by relationships, where happiness is fulfilled by relationships, and where language in words produces the growth of knowledge of the world in which we find ourselves.

Patient HM—A Person of Interest

The musician Prince, who was born Prince Rogers Nelson, was known for much of his career as just Prince. Then in 1992 he changed his name to a symbol that could not be pronounced or reproduced with any computer font. It was both a contractual tool and a symbol he later revealed was for love. When anyone referred to Prince in speaking or printing they had to say "the person formerly known as Prince." And that name stuck.

But these few paragraphs are not about the musician formerly known as Prince. The person I want to mention is best referred to as "the person formerly known as patient HM." In his unusual case HM's personal experience was lost, not the person, but the experience of being a person, because he disappeared into permanent present tense. HM did not disappear into a coma, but into a freezing out of time. Because of a need for surgery for epilepsy, HM lost his ability to make new memories and remained in present tense for the rest of his life.

In August, 1953, Henry Molaison, a 27-year-old man, had surgery (a bilateral medial temporal lobectomy) to remove his hippocampus, to relieve his severe epilepsy, which it accomplished. But no one knew what the removal of the anterior of two thirds of his hippocampus would do to his memory. Henry immediately became a "minute man" (his memory lasting usually only minutes) and in the scientific literature he was called patient HM, for his privacy, until he died in December of 2008, at the age of 82. Unknown to anyone at the time of the surgery, the hippocampus was the center in the brain for consolidating memories from short term to long term. Short-term memories may for the most part last anywhere from 30 seconds to 30 minutes. Most are less than two minutes long. If someone tells you her phone number and you go on to other things, you will have forgotten it in several minutes unless you made a special effort to remember it.

The devastating problem with this loss for HM was that he could not have any new memories (and he actually lost many years of past memories as well) and remained stuck in present tense. You are always asking your doctor what her name is, and even years later you still do not know her or her name. HM, looking in a mirror many years later could not see the young man he remembered, just as Oliver Sacks reported with his patient Jimmie G. who also had lost his memory forming ability. Eventually Jimmie stopped looking into the mirror. The result is that without forming new memories we always feel as if we are waking up and living in minutes that are all alone in themselves, not connected to any other minutes, as if no other time exists except the immediate moment and the very distant past. We all exist in present tense, but for us our experiences are connected to the past and the anticipated and planned future. That connection gives us our sense of self.

After HM's surgery and its destructive aftermath, he became one of the most important people in the research on the brain and memory. It is nice to know that you will always have a job I suppose, and if you did not like it, you would not remember what you did not like about it. HM spent 55 years being one of the most famous subjects in the study of the brain, largely at MIT, but he would not remember a day of it. At his death his name was revealed, having been HM all those years to protect his identity. After his case, where HM had been mentioned in over 12,000 articles in medical and psychological research, he was just an unknown, or only known by his initials. Henry Molaison also suffered a retrograde memory loss of eleven years before his surgery, back to age sixteen, so he never really aged a day since age sixteen. He did learn and retained new motor skills, although he never remembered learning them. After his death, his brain was dissected into 2000 slices and digitized as a three dimensional brain map that could be searched by narrowing down from whole brain to individual neurons.

We cannot know exactly what it would have been like to be in present tense, never existing except for a minute, and then quickly gone again into the next minute. Maybe the experience is like when you are roused from your sleep, and groggy for a minute, not knowing what day it is or where you are. But make no doubt, Henry was still a person, but excluded by a brain without two thirds of his hippocampus on both sides of the limbic system from collecting himself together over time and feeling like a self. He was lost, but maybe no more than any of us in the grander scheme of things. We certainly would benefit from trying to gaze into his mind to see him instead of just finding a place for him to sit and wait.

Seeing the Hemi Neglect Person—Oliver Sacks

There is a strange malady called hemi neglect, also known as unilateral spatial neglect or unilateral inattention, where a person does not recognize or admit to part of his visual field, or part of his body. The disorder can take many forms and from damage to several brain areas. However the most common form is left visual hemi neglect, caused by damage to the right parietal lobe of the brain. The parietal lobes of the brain can attend to the space on each side of us. The hemi neglect person may eat the food only on the right side of his plate and not even consider that there is half a plate he is not seeing. One hemi neglect patient cannot read easily because he can only see the right side of the page or the right half of a word. Such persons may refuse to talk to anyone standing in the left field of vision, or to admit that the left leg or arm belongs to them. These patients seem to have lost the motivation or the ability to attend to things or to respond to things on that side. It is not a visual problem, but a problem that has to do much more with the overall consciousness of the person. These people are not purposely trying to be difficult. This is the way they are conscious of reality.

Hemi neglect occurs in many forms and possibly damage in many brain areas. Sometimes hemi neglect patients will even disown a left arm or leg on their bodies. It is not a question of failing to see the arm or leg, but a lack of awareness of even the existence of the limb or a problem or no willingness to mention it. A famous example of disowning a leg is from Oliver Sacks, who badly broke his leg when being chased by a bull on an uninhabited mountain in Norway (*A Leg To Stand On*, by Oliver Sacks). In the hospital after the surgery to set the leg, he would look at it, his leg, and argue that it was not a real leg at all, or it certainly was not his leg. He would touch it and he felt nothing, it was alien to him, a lifeless replica attached to his body. He also could not identify with or feel the stitches being taken out of his leg.

There is a problem called Body Integrity and Identity Disorder (BIID), where a person has a strong desire to remove one of his own limbs. It feels intrusive to them, unwanted. The cause may be an abnormal amount of activity in the right superior parietal lobe. Here is Oliver Sacks's reaction to seeing his left leg in the bed after surgery, "… I raised myself on one arm, and took a long, long look at the leg… A leg—and yet, not a leg: there was something all wrong. I was profoundly reassured, and at the same time disquieted, shocked to the depths. For though it was 'there'—it was not really there."[1] To Sacks the leg, his leg, was an alien leg, attached to him and in bed with him.

The Locked In Syndrome—Trapped in a World of No Body

In Santa Jose, California, in the summer of 1982, Dr. J. William Langston, chief neurologist at Santa Clara Valley Medical Center came across four cases of almost completely paralyzed individuals suffering from what appeared to be Parkinsonian paralysis. It turned out that they had become paralyzed in a matter of days after taking a street version of the drug heroin. The alarm went out that bad heroin was being sold on the street. The street version, cooked too long, made MPTP instead of MPPP. A small change it seems, but the change destroyed their substantia nigra cells in the brain, which produce the transmitter dopamine. A small chemical change produced hugely horrible results. They became instant Parkinson's patients, perfectly healthy patients with active minds locked in almost completely paralyzed bodies.

A similar situation occurred much earlier and involved patients paralyzed for many years by the Great Sleeping Sickness epidemic of the 1920s. Oliver Sacks treated and awakened these unfortunate individuals, but that could only be for a limited time before the L-dopa, a precursor drug to dopamine, had no dopamine-producing neurons to turn it into dopamine.[2] How sad to a healthy mind that is. The disaster did not just leave the patients paralyzed but affected every area of their conscious life including thoughts, perceptions, and feelings. Everything was slowed down to a dreadful crawl in the Parkinsonian process. Also, frequently common in these patients, at the same time, were catatonic disorders of many types, including maintaining various postures indefinitely, echoing thoughts or words that had been suggested to them, and tics of busting out of catatonic states into spontaneous, uncontrolled movements.

Persons are still persons no matter what the state of their brains and their abilities to show their individual personhood. Some are far out on the edge in comas and might recover like Terry Wallis of Arkansas, who woke up 19 years after his car accident in 1984 left him in a vegetative state. Some persons are gradually moving toward the edges of personality, further away from us every day, and will not recover at least with today's medicine. Such persons are lost in Alzheimer's disease, sometimes called the walking dead because they will not recover. Millions of Americans now have Alzheimer's disease and fifty percent of Americans over 85 will get it. And if we do not get it, we will be taking care of someone who has Alzheimer's. Some persons are on the far out edge of personality, already lost except for being kept alive in what is called the vegetative state, occasionally awake but without conscious awareness, and the body carries on without the mind. Recent research with brain scans raises the possibility that many of these vegetative patients are conscious at some level and can communicate and respond to commands or yes or no questions. In summary, all human beings are persons in spite of their difficulties, and should be treated accordingly. Maybe that view point could guide our research and treatment of them.

Concluding Thoughts

At the height of the racial tensions of the 1960s in the United States, two individuals, one man and one woman, separately researched what it was like to be a black person living in the racially divided south. Their research was not from the scientific literature or surveys or newspaper accounts, but as whites becoming blacks and traveling throughout the south. They wanted to know through personal experience the answer to the question what a black man or a black woman experienced in our racially prejudiced country. They both used medication to darken their skin. Then they exposed themselves to sun light to complete the effect. The intent was to not just observe, but to experience the lives of blacks in a white world.

John Howard Griffin wrote the story of his journey through the Deep South in the book *Black Like Me*, in which he recounted his travels of six weeks through the four racially segregated states of Louisiana, Mississippi, Alabama, and Georgia. With his temporarily darkened skin he could pass as a black man. In a similar effort the book, *Soul Sister*, written by Grace

Halsell, who also blackened her skin after the assassination of Martin Luther King, Jr., journeyed as a black woman to the Mississippi Delta. By literally putting on the skin of a black person, they each reported seeing from another person's perspective, both rationally and emotionally, and thereby coming to a deeper understanding of persons in the racial divide between whites and blacks in this country.

The thought behind all of these examples is that you do not learn what it is like to be a person with just experimental research, although that is important, but by remembering that personhood is part of the data when studying the brain in a person's head. Our knowledge of the people with brain disturbances should be deeper than just statistical information or biological research. We may need to understand persons better in order to better treat their brain disorders by at least searching out and really seeing the essence of their condition as real persons with blocked avenues to themselves and others. As far as treatment goes, if it is relationships that are so important to human beings, then it is relationships that many brain damaged patients lack and that we can give. Such patients are still persons after all.

People in wheelchairs often say they are ignored by people as they try to shop. People just tend to look over their heads because they are not adult height. The reason for that may be that we as "normal" people are too embarrassed to deal with the reality of paralyzed individuals. It is almost as if they are second-class people. At the time of the summer Olympics I realized that in a world of Olympic Stars, who could all run the 100 meter dash in under ten seconds, I would seem like a handicapped oddity. In a world where the average IQ was 240, I would seem to be profoundly retarded. In a world of IQs of 80, I would be a genius. Personhood is not what I can do or say, but who I am in God's view of and love of us.

Our personalities are all fractured in one way or another, and we have to take into account that the world is not filled with two kinds of people, the beautiful and the ugly, or the "sick" and the healthy. We need to see others and ourselves as persons, with some just needing to be understood and seen, and some needing to be healed. We all need to see all human beings as persons and let some personal information be the clues to how to treat those on the edges of normal. If we take this guide of personhood first, and never stop looking through those "personal" glasses, then we will do better in our treatment of, and our research on, those we work with in the medical field dealing with brain trauma.

We are always persons, even if we cannot speak, or are in comas, or cannot do anything that is human. Those who believe in the immortal life of a human being can see human life as more than matter or function. And we all need to choose to see that there is more to a person than physical life. We do not require for there to be self-awareness, choice, language, thinking, and relationships to be present at the beginning of life to know we have a little person in our arms. So too, we have not lost a person when massive brain damage has removed those same things. We see personhood extending beyond the boundaries of the brain and time and into the resurrection of the body into eternity.

Christians believe that is what God's relationship with believers is. The Holy Spirit inhabits the heart of each believer. God has entered the world with human nature, and then enters each human who believes and trusts in Him. What I realize now since writing this chapter is that He really sees and understand me better, person to person. This is the relationship desired by God with human beings, not to be just church attenders, but for us to relate personally and individually with the God of the universe, and each other for all eternity.

***

> **"We're trying to shove square brains through an oval world."**
>
> **—Diane Ackerman (*An Alchemy of Mind*)**

SOME BOOKS I THINK YOU WOULD LIKE

Thinking in Pictures: My Life with Autism, **by Temple Grandin.** This brilliant PhD has transformed the world of animal science by redesigning animal handling facilities all over the United States. She sees and thinks, not in words, but in pictures, a concept of perception not easily understandable to the rest of us.

An Anthropologist on Mars, **by Oliver Sacks.** This book is Oliver Sacks's medical case record of seven individuals with neurological disorders. He leaves his doctor's office and joins them in their daily routines, such as the doctor who has severe muscle spasms because of Tourette's syndrome, except when he is doing surgery.

CHAPTER 8

THE MOUSE-HUMAN BRAIN

There are brains on the planet that are part man and part mouse. However, there is no need to fear an alien sci-fi attack by killer rodents from these mice. Neuroscientists have merely added human glial progenitor cells to newborn mice brains, producing whole mice with a small percentage of human brain cells. As the mice with these human brain cells matured, their memory and learning skills were better than ordinary mice. One purpose of adding human brain cells to mice brains was to eventually end up with animal brains similar enough to human brains such that these new brains could act as living laboratories for research on human brain diseases, such as Parkinson's or Alzheimer's. Scientists cannot ethically experiment on human beings in ways that they could with animals whose brains were more human like. Give the mice Alzheimer's or Parkinson's and then work on the cures. Thus, we could have a living-laboratory animal brain, and no humans would be harmed. Sheep have brains that are closer to human size than mice, and their brains could act as such living laboratories. However, no medical researcher wants a pen full of "bahha-ing" sheep down the hall in the science lab, right? Mice, therefore, have been the experimental animal of choice.

A TITLE CAN SAY A LOT

Super Brain

—Chopra and Tanzi

RECREATING THE HUMAN BEING: FUTURE NEURO-TECHNOLOGIES AND ROBOTS

"O brave new world that has such people in't."
—William Shakespeare (*The Tempest*)

Dr. Frankenstein worked on his creation, madly and tirelessly, night after night in his laboratory. He was sewing together a man from dead body parts, and then he would draw its new life from electricity. Seems like a good idea even today when you think of delivering electric shock to the heart attack victim's chest in order to restore life. When Dr. Frankenstein finished his surgical work, he looked at the creature, and he hated and rejected the ugliness of his creation. Then, the monster (whom we call Frankenstein) escaped and gave Mary Shelley a great novel. Ignoring most of the later movies, the monster Frankenstein was not all that bad. Except for his murderous feelings of revenge, I suppose you could even feel sorry for him at times.

Some readers of *Frankenstein* saw God as the evil creator, who made man imperfect and then left him alone to fend for himself in an unforgiving world. In the words of Milton's *Paradise Lost*, "Did I request thee, Maker, from my clay to mould me man?" Some saw in the novel a failure of science that was represented by Dr. Frankenstein and any experiments going on at the time that were attempting to bring animal corpses back to life with electricity. Other readers could see themselves in the monster's role. Who has not felt ugly at times, hopeless and without friends who cared? Think of the other sad, ugly monsters in literature from the Hunchback of Notre Dame, Cyrano de Bergerac with his big nose, and The Elephant Man of real life, all of whom were rejected because of their ugliness. In every interpretation of Frankenstein in this view, the monster is almost always us. The monster cried out in his agony in Mary Shelley's book, "I was cursed by some devil, and carried about with me my eternal hell." We would do well to be careful of our future considering some of the amazing tools for changing human beings coming our way soon. Are we trying to become like a god in our scientific power, or are we pursing a moral God-likeness in character? Perhaps the more important question is, have we stopped thinking of human beings as persons and have just gotten too used to the idea that we are all machines to be reconstructed for the future?

The question about what exactly is a person arises when we create mice or chimpanzees with increasing percentages of human brain cells. If we wanted a higher percentage of human neurons or glial cells in the mouse brain, we could put more human stem cells in the newborn mouse brain. With two percent human brain cells in mice, the mice get smarter on the maze testing. Therefore, the limit was put at 1% human brain cells in order to be ethical about what exactly these mouse-human brains were. The human neurons and glial cells were longer in size and more extensive in their spread throughout the mouse brain. The only thing that had to be done to have a mouse with 100 percent human cells would be to replace all of the mouse neural cells with human neurons in the growing period of the mouse brain. The mouse would have a human brain, but probably not the architecture of a human brain, meaning a large frontal lobe and a speech center. We could almost as easily produce a chimpanzee with 100 percent human brain cells, a chimp with a human brain! It is clear to see why such experiments have been forbidden in American research, but someone is undoubtedly upping the percentage of human cells in animal brains elsewhere in the world. Our world would then become the world of the chimera, the mixture of animal and man, and the oh-so-strange world of H.G. Wells's 1896 novel, *The Island of Dr. Moreau*. Greek mythology described the chimera as a monster with the body parts of the goat, lion, and a serpent. In the same way some predict we will eventually see the mixture of man and animal, or man and computer as soon as the need and the expertise arises.

This emerging world of neuroscientific, technical advances on the human brain is not just science to help us see better, walk better, and do our daily chores better, but it seems to be easing toward the attempt to create consciousness—the Holy Grail of neuroscience research, to finally crack open the hard problem and see its empty center. Such is the hope. To create consciousness in a machine some say will show the evidence for the emergence of the human consciousness from putting the pieces of brain together. The whole is more than the sum of the parts you know. Out of matter will spring mind or soul, supposedly since there is no God, only pieces of matter combined in just the right way. The mysterious matter of mind will not be so mysterious after all. The very act of creating life and mind makes us a bit more like God, some suppose. If materialistic assumptions are wrong, I fear that we will find ourselves all weary in Dr. Frankenstein's lab late at night with horror and fear at what went wrong in our remaking of human beings. Even if no harm should come from all of our research in these directions, have we just been working on the simpler aspects of the neurophysiology of humankind and ignoring the human essence, personhood, which also needs our attention?

The lesson or theme here in this chapter is to, in these times of scientific advancement and change, be even more aware of the personal in the mystery of our brains, where meaning, significance, and purpose in life become important. We also need to consider how the fallenness of our natures, and our spiritual needs and growth are being dealt with in those areas of ourselves ignored by our materialistic culture. This was the problem when we humans moved from covered wagons and horses, to autos and planes and space ships, only now everything is moving faster, fueled by profit and worldwide information, and a spiritual and personal shallowness all around us.

What Lies Ahead

Our future in the science of the brain looks amazing, but it should also make us pause. Pause does not mean stop. It means to think for a moment. Think along the lines of becoming what we humans ought to be. Neuroscience has gotten to the point of possessing the ability to remake aspects of human nature, and that is where our serious thinking and ethical work must start, with the defending of ourselves as persons. If we start with man the machine, then we will find only machine ethics and outcomes. The science of man is ready to take over from the ever-so-slow methods of survival of the fittest. It is said that we, with our activity scanners, recording electrodes, and genetic tools have begun already to take over our own evolution.

The immediate future may possibly hold mind-to-mind communication, the merging of human and computer minds, a brain-net to mentally read texts, immersion entertainment to not only see but also feel the emotions and excitement of the entertainers in rock concerts and movies. There is every reason to think that in a few years all of these will become a widespread reality. Military applications of these emerging sciences are numerous, from increasing the sensory abilities of soldiers, guards remaining vigilant without sleep, chemicals to make prisoners reveal secrets, manipulating human minds even at a great distance, and increased sophistication in brain-machine interfaces. Newly developed computer chips will allow human brains to mentally see digitally flagged targets in battle. Drugs are being developed that will erase bad memories or prevent posttraumatic stress disorder (PTSD) from the horrors of war.

Other advances in neuroscience show it is possible to insert in animal brains basic memories using brain chips and computers. Human experiments have been done with computers and EEG readings to record and print pieces of dreams, and someday you may be able to buy pleasant dream starters to use for your own night time dream entertainment. The peaceful EEG waves of someone's healthy brain may soon be inserted via a transcranial magnetic stimulator into the depressed or anxiety brain for improved mental health. Even now one can purchase headsets off the internet that use electrical or magnetic pulses to stimulate or quiet various areas of the brain in order to improve the brain's function or help treat depression and anxiety. A lot of research will go into enhancing the brain, making it better in every way possible. Electrical and magnetic stimulation of the brain will be used to enhance cognitive and athletic functions of people. Improving brain function will be speeded by the new research showing that skin cells can be transformed into nerve cells and used to repair, replace, or add to brain areas. Lab grown spare parts for brains are now being planned as well.

Lie-detection devices will be perfected. Brain-stimulation research is now being developed to improve creativity and possibly generate more genius-level abilities in all human brains. New and developing techniques in animal research include the technical ability to move simple animal memories from the hippocampus of a mouse and store those memories in a computer. Later the researcher will input those memories into the brain of another mouse. This seems very possible to do in human brains with simple memories, which you wish to erase or create. The results of robots with false past memories are seen in the film *Blade Runner* with actor Harrison Ford, where only the expertly trained could tell who was a robot. The humanoid robots with memories of their false past did not even know they were robots.

Imagine music concerts with full immersion into the mind and emotions of the rock artists as the connection is made with EEG-like devices on the sending end, and the direct transmission of those waves into your brain on the receiving end via the internet. The same would be true in some ways with a totally mental Twitter, Facebook, and texting through a world-wide brain net. The ultimate goal could become for many to achieve immortality itself through downloading your present mental state and memories into the permanent hardware of a computer, and to exist as long as that software of you can be transferred time and time again. Talking about the possible consciousness of computers and robots suddenly becomes an important matter to discuss.

However, a few years from now what will we think of our re-creation of ourselves? And which of our best plans for mice and men will actually work the way we hope they will. There will be much bionic good for the lame, deaf and blind, and even the Alzheimer's or schizophrenic patient. But what will we think of the new culture, the new morality, and the new man we will be able to produce with our new technology? Now is the time, not tomorrow, to think about the directions we take, and the persons and culture we want to produce. Many of these advances can be good. And many will not be. It is good to remember that what interests us in our culture in terms of emerging technologies, from computers to iPhones grabs not just our wallets, but also our time and attention, and ends up possibly shaping the plasticity of our brains to fit the new culture.

These future technologies are intended for both the enabling and enhancing of human brains, to both aid the handicapped body or the diseased brain, or to vastly increase the human abilities both mentally and physically. Ethical discussions will hardly keep up with all the possibilities of a neural society unless we adopt a person-oriented point of view in our neuroscientific explorations and applications that can give us a limited number of general, person-oriented guidelines. It would be for example only such guidelines which would remind us of the ever-present dangers of increasing the gap in our world between the haves and the have nots in society, those who have the money and the education, and those who do not have the money or access to all that is new in science.

It is all of us who are at stake in this technical brain revolution. Angelika Dimoka, director of the Center for Neural Decision Making at Temple University, used fMRI measures in the dorsolateral prefrontal cortex as volunteers were subjected to increasing amounts of information overload. She found that activity in this frontal area of the brain fell off, mirroring the drop in decision-making accuracy and the loss of emotional control of her subjects.[1] We are all subjected to such loss in our busy, techy culture and we do well to cautiously examine the future developments.

I will now cover several categories of neuro tech advancement, which will illustrate the coming age for us, and allow us to think more holistically and personally about humans in the future. Some advances are obviously needed and helpful, and others make you wonder who is steering the scientific spaceship. I am focusing on the kinds of things we might see in the near future, such as using thought for remote control, the networked mind, the new mental health approaches, and the coming of the robot into human lives and culture. Remember that your view of the brain/mind question necessarily leads to what you believe about the sources of human problems and where you look for answers. This is not the day or age to surrender our Christian view of the person, but to keep personhood in mind lest we lose the future of the human person.

Our Minds as Remote Control

Jose Delgado, a neuroscientist at Yale, bravely entered a bull ring in Spain in 1965, armed with only his wireless transmitter, to fight a charging bull.[2] Delgado pressed a button and the signals went out to electrodes implanted in the bull's brain, and their signal stopped the bull's charge. The bull was not changed into a peaceful bull, but it did stop. As mind/brain control becomes more common in the future you may actually become the remote control for your garage door or your television set or the lights in your house.

You may have I-skin attached to your forearm and just your thoughts will send signals to perform functions. This flexible artificial skin attachment to your arm has sensors with nearly invisible touch controls and wireless transmitters. In ways similar to the quadriplegic controlling a computer cursor, you will think a certain thought or flex a certain muscle, and your brain will send a signal to a wireless transmitter which will turn on the lights in the room or start your car, or even move your paralyzed legs or artificial limbs. A paralyzed subject's thoughts can be used in a similar way to change the television channel. The subjects are not thinking channel 2 or 16, but the television iPlayer cycles through the channels as the subjects concentrate hard for the channel of their choice. Strength of concentration makes the choice. For something like opening your garage or turning on the lights, only one choice is needed, so any thought would do.

Non-invasive brain-computer interfaces are being used to assist the paralyzed person not only in walking but also in controlling a computer cursor. With an electrode cap or a nearly invisible strap slipped onto the head of the paralyzed person, the EEG brain waves during thoughts will be picked up and sent to a computer. Different brain signals are then tied to different commands that the person needs to execute on his wheelchair or computer. Newer technology from a company called Ossur in Iceland allows a person with a bionic leg to send thoughts about moving his legs, since electrical signals still exist for such thoughts through the thigh muscle neurons to control the leg for walking. It works so well that subjects can perform the actions with only a little training. The action is from thought to muscles to sensors to prosthetic leg movement. Even current prosthetic legs can be updated to work with mind control.

Paralyzed individuals also have been testing control of a robot remotely just using their thoughts. This not only allows them to control a robot just as they can control a computer cursor, or a leg, or a television, but if a camera is mounted on the robot and it can move around a house or school they cannot be physically present in, they can feel like an avatar, actually there, and not just in their bed at home. The average days of training it takes to control the avatar robot as it moves around is about 10 days. The practical applications of such research findings have been used by test pilots wearing EEG recording caps studded with electrodes to fly and land aircraft in test simulators using only their thoughts. Nothing seems wrong with this type of control, especially when we see the possibilities for the lame to enter into society again.

The Networked Mind

The digital linking of live animal brains together in a common network in what is in effect the creation of a "brain net" was demonstrated by Duke University neuroscientists recently. The scientists linked the brains of rhesus macaque monkeys to perform simple tasks. The animals had a shared "hive" network of sensory and motor information, and with their collective brains they were able to solve problems better than a single animal alone. They dubbed this network a "super brain" made up of three monkey brains. The experiment required the cooperation of at least two of the monkeys in order to move an avatar's arm. The future can include the sharing of more complex information such as personal and emotional information between people. It is not too difficult to think of future problems with this application, but maybe I am just remembering too many Star Trek episodes with the powerful life form the Borg, which had locked many unwilling minds in place for its attack on the Starship Enterprise and the rest of humanity.

The research that drew wide attention recently in neuroscientific advances was the first human-to-human brain contact, from the University of Washington, called the "Vulcan Mind Meld" after an ability shown by Star Trek's Mr. Spock and his Vulcan relatives. This demonstration of human telepathy was achieved as the brain researchers from the University of Washington used brain waves from one person to control the finger motions of another person. In this case EEG waves taken from subject number one were fed into the internet to the computer of subject number two, which then activated a transcranial magnetic stimulator resting on the head of subject two. When subject one merely thought about moving a finger to fire a cannon during the video game, then that thought affected his brain waves, that information was transferred into the internet to the computer of subject two, which then activated the magnetic stimulator mounted on a cap on subject two's head. The magnetic pulse then caused the electrical activity in his brain to move his cannon-firing finger. This is mind-to-mind control. It is not a big deal actually, but it is seen as the beginning of more minds entering other minds to communicate. The distance between the two minds does not matter since the internet fills in the distance. Should we be worried about someone controlling us by such brain control? That possibility depends on whether people have a choice on whether they will enter into such controlling and group think situations.

Minds Open to Other Minds

The ability to use brain wave data from a person and transfer it to a computer and then use that information to produce similar neural activity in another person's brain is now being developed. Computers are able to reasonably reconstruct from brain activity the images of faces that subjects were viewing. None of these studies shows the transfer of real thoughts from person to person. But they do show the possibility of transferring information, emotions, or thoughts from one brain to another in the future, using EEG recording on one end and TMS (transcranial magnetic stimulation) or tDCS (much cheaper) on the other. And perhaps the ALS patient or the completely paralyzed person will be able to use a brain-to-machine device such as a voice synthesizer just with his thoughts. Military uses of covert operations requiring communication under the cover of silence are also envisioned, as well as making use of stored memories of physical abilities like dance or piano, to be placed in another person's brain. Additional scientific possibilities include tapping into the minds of chimpanzees or dolphins to answer questions about their thinking and awareness abilities. Perhaps when you think of a new idea or imagine a possibility of some new invention, your computer could collect and store your 3-D film of what you imagined. That's better than your present note pad or iPad that you presently pull out on which to jot down your thoughts.

To read the emotions of others in the future is becoming a real possibility. Many of us in childhood have bought the mood rings that would change colors when your emotions changed, red for warm toward a person, green for friend, and blue for cold toward someone. Carnegie Mellon researchers recently demonstrated a modern, and certainly a better, version of the mood reading ring, which involves researchers asking method actors whose brains were being scanned to generate various emotions.

Sure enough, the moods of fellow actors could be accurately guessed using just the brain scan data. Similarly researchers in Korea have used Band-Aid-like goosebump sensors that can accurately monitor strong emotions such as fear or pleasure. Accuracy wise, the use of these methods as lie detectors is currently debatable. And if information is all we want, then fine, but the experience of another person's emotions in our emotion-to-emotion contact is far better and more human.

Some other future possibilities include, with commercial thoughts in mind (isn't that what often drives research?), looking into your brain and seeing your dreams. Japanese neuroscientists have used scanning technology to see rough versions of what people are dreaming. The pictures are sketchy, but they will get better. The plan is that we in the near future are able to download your EEG waves from you dreams, not for better psychiatric counseling, but for sale. Your EEG waves will have already been catalogued for thousands of pictures and social and action situations, such as walking upstairs, talking to a person, running away, and so on. Research shows that the computer can then draw, from the EEG information, what you are dreaming. When the process is reversed, it is hoped, we could download someone's dream EEGs using a magnetic stimulator into someone else's brain, and let the receiver's brain pick up the story line to run with and to embellish the exciting dream content. You could buy an exciting dream from a dream factory like you buy DVDs or stream movies on to your television today. Or you could replay one of your own favorite dreams over and over again. Your computer will be attached to your playback headset wirelessly, and off to sleep and pleasant dreams you are.

The new technologies might also mean that you could download the general emotions and vague thoughts of the latest rock singer's concert you are listening to and watching on your computer screen. The singer at a concert in France would be wearing a head band that would pick up her appropriate EEGs paralleling her emotions during the song, and that EEG information would be sent by internet to your computer to your transcranial magnetic stimulator to create the same brain waves in your brain as you watch and listen to the concert. You more realistically could sense and feel the wildness, the emotional ups and downs, the sensuousness, and the freedom of the singer. This will be a part of the new immersion entertainment of movies and television, where all five senses will be stimulated electronically in the brain corresponding to the actors and the drama unfolding on the stage. We will be living in the avatar world presented to us through our brain connections in much the same way shown by the movies "Avatar" and "The Matrix." Evaluating the entertainment of teenagers by concerned parents will be much more difficult.

It is reasonable to expect that the Twitter of the near future will be carried over what is called the brain-net internet (an internet in your head) and you will be enabled to enter wirelessly into certain people's brains.

This all seems more possible since scientists in Sweden have built the first artificial neuron that can mimic even the chemical functions of living, organic neurons. From Harvard University injectable brain implants have been developed. They are small enough and flexible enough to roll up and be injected into the brain with a syringe through a small hole in the skull, and unroll after being injected into the brain. These mesh brain implants can then function at the level of individual neurons. It is conceivable that you could receive news reports, friend's postings, stock market quotes, songs, and mental phone messages. Someday, maybe, you could arrange group meetings at business and participate in everyone's thoughts. You could sit and listen to a worldwide conference on global warming without even leaving your home. Your children would be just a "thought" away. Of course, many of these options can be done with the cell phone and radio right now. So, why try anything else? These devices will be better than present technology since they would literally be putting you "there." Research on all of these technologies is being done today. One should think that all of these possibilities are at least fifty years off in the future, but our experience suggests that if we want it or if it sells in our entertainment-minded industries, it is closer than we think. And our past experience tells us that we are not even capable of thinking of all the possibilities that are rapidly becoming realities.

The Robots Are Coming

In July of 2016, it was reported that a mall security robot, which was patrolling a shopping center in Silicon Valley, ran over a sixteen-month-old toddler. The boy was not very injured by the five-foot, 300-pound robot on wheels that looked like a friendly egg-shaped soft drink machine. These security robots are programmed to alert human guards during any loud noises or changes in their environment as they wheel around the mall. These kinds of accidents or worse are only going to happen more in our robotic futures. We cannot see the future of robots, but many who fear it say we should begin to think about it now.

No one author has popularized a future with robots as much as author Isaac Asimov and his sci-fi Robot books and his Foundation series of books. Asimov early introduced his Three Laws of Robotics to explain how we would never be threatened by robot intelligence or power. All robots would be programmed according to the following laws.

1. A robot may not injure a human being or, through inaction, allow a human being to come to harm.
2. A robot must obey the orders given it by human beings except where such orders would conflict with the First Law.
3. A robot must protect its own existence as long as such protection does not conflict with the First or Second Laws.

These Three Laws have been referred to in many science fiction books and films over the years, and they help answer some questions about highly intelligent and powerful robots, and the safety of human beings around them. Much has been made of the "singularity" by computer scientists and others about the point in the future when computers would be smarter than human beings and would be totally in charge of their own programming to improve themselves. So the question becomes, will humans be in trouble when that time comes? And will the machinery and electronics of robots ever produce consciousness, and make robots partners with us in life on earth? Many researchers answer yes to both questions. Many almost seem to hope so, thus, removing any need for the outside-of-nature creator God to explain things.

Robots will work for us, and perhaps do most of our work for us, from the sweeper-on-wheels in our homes moving around the floor all day, to the humanoid robots that we may not be able to tell that they are robots. Most calls to customer service phones will be to robots who have been programmed, not just to respond with information, but to respond with emotion and empathy in their voices. In the future these robots with programming for empathy are going to be quick with humor, emotional language, and changes of attention to follow human discussions. We may decide that robots cannot be conscious, but these will eventually sound and act as if they are.

Pascale Fung, professor of electronic and computer engineering at Hong Kong University of Science and Technology, says that such emotional robotics must do more than ask, "Can I help you?" They must be attentive to the acoustic cues of stress, anger, happiness, boredom, found in the speed of conversation and the changing of pitch in our voices, as well as understanding the meaning and intent of human speech, and the context that might change the robot's responses. We do this so naturally, but it will be quite a learning curve for a robotic machine. The progress of robotics suggests that robots will be developed sufficiently to pose as human beings. Robots do not have to be perfect to be a part of our world. In Japan female robots have begun to work at a department store, and in Singapore, the robot Nadine is a receptionist at Singapore's Nanyang Technological University. These robots are humanoid in appearance, and show personality, intelligence, and emotions. The fact that we can build robots that can be cashiers who can speak in several languages as they attend to the needs of shoppers, means that we may have to discuss the rights of robots as well as the rights of the people who are displaced by supposedly machines in the work place.

Will Robots Ever Become Conscious?

The Turing Test, developed by Alan Turing in 1950, was supposed to see if artificial intelligence was indistinguishable from human intelligence, the Holy Grail of every AI researcher. Will a computer of the future be so good that we could not tell whether it was human or not—*i.e.*, we could not tell if it was just a machine with no consciousness inside the machinery? Turing said at that time a robot would deserve to be called human and to be treated as a human. But before we think that robots will easily demonstrate human intelligence and personality, there has been developed a tougher Turing Test to compare computer and human thought and intelligence. The new test is called the Winograd Schema Challenge. The new test, developed by Hector Levesque from the University of Toronto, contains sentences with language that would be ambiguous to a robot, but not to a human, since the sentences require common sense to understand and not just a good dictionary. So far, we can still best the robot with much of what we can think, even if we lose at chess, and speed and memory tasks. Common sense requires putting many bits and pieces of information together and the ability to think from them in new and sometimes unusual ways.

There is no machine equation to confirm something as self-evident or common sense to a 6-year-old or a grumpy adult. You can read a child or an adult with just a second of thought, but it is not that easy for a computer brain. Computers cannot plot realistic simulations of the future, although plotting the future in chess is done by raw computer power in a world where all possibilities can be known. The efforts of AI researchers are moving forward to make computers more like humans with neural networks, collections of neurons that constantly reorganize themselves after learning something new. The human brain has no central processor, but its networks are massively parallel with all of its neurons and glial cells working simultaneously in order to immediately respond to the environment, and to be learning at the same time. Researchers are now developing the parts for brain-like computers and so the possibilities of building more human-like robots is growing. Human brains are fast, have almost limitless memory, create very little heat, and require just hamburgers to power. In 2014 scientists unveiled neuromorphic chips designed to handle information in parallel like human brains work. There is still a long way to go in order to produce a computer brain like ours, but the beginning steps are being made.

For some, progress will be made when human and machine brain parts are combined together in what is called a cybernetic organism, or cyborg. This was first demonstrated in a sophisticated way with a robot navigating around on a floor maze controlled by 300,000 rat brain neurons. Another example was the I-Cub Grow Bots that play with and learn from children by gaze tracking (following a child's gaze) and imitating children's words and behaviors. The I-Cub revises its behavior in terms of trial and error and makes progress. But it cannot right now make the progress children make in language and the recognition of faces. An ethical question will soon be asked about adding human neurons and glial cells to such computer devices. Then we will see how the abilities of such cyborg computers will be ramped up.

To upload our entire selves into a computer and be all digital is another question. Merely copying all of our NMDA memory receptors, or whatever engram we have decided is the physical basis for memory, will not be us.[3] A set of memories is not the one who uses the memories, who builds self and self-concept on those memories. A book on the shelf is not the author, just her thoughts. A video of a man speaking of his experiences is not the person. It is just a memory bank. And even most of our daily memories are forgotten anyway, when our episodic memories disappear in favor of the meaning of those periods of time and thus our personhood is rooted more in meaning than in objective memory engrams.

Mental Health and Brain Technology

I will not cover all of the many mental health areas that are being transformed by our knowledge of the brain and its operation, from drug addiction, sleep disorders, Parkinson's, pain management, Adverse Childhood Experiences (ACE), and more. I am only picking a sample that will make my point, which is to not forget the personal part of us in everything we are and do. Let our future scientific developments spring from a view of personhood, and from our deepest needs as persons, in particular, finding purpose, meaning, significance, and fulfilling relationships.

Dr. I Robot—Counselor

Developing robots as mental-health counselors is already being considered. Sometimes we find it consoling to talk to a nice voice even if we know it is a robot programmed to answer all questions well and patiently. Siri or Google seem already to be heading in that direction. Robots fully indistinguishable from humans, full of emotions like kindness and caring, may not be humans, but they could certainly be programmed to appear to us that way. One of the biggest programming hurdles is going to be programming the empathy "molecule" that will recognize the facial gestures, tonal changes, body language, along with the message in the speech itself. Once the robot can accomplish that, then it is a matter of deciding what to do and say in order to help the person seeking counsel. Once we cannot tell the difference between the robotic counselor and the real thing, then, the Turing test would say that the robots are in effect caring, human-like beings no matter what the unknown state of their minds. Most human counseling problems boil down to just a few dozen, if that many, and the same answers can satisfy the clients, over and over again. Kindness and empathy, being a person to a person, is the best therapy, and computers could manufacture that without being persons.

Thought-Controlled Mood Medicine

Research from bioengineers in Switzerland has shown that human beings, by thought alone, can control a device implanted in mouse skin, and turn on certain genes in that mouse. The design is the familiar using human brain waves to control an external device using a brain-computer interface, and the device is used to turn on a fiber-optic switch controlling neurons or genes themselves. While this is many years from affecting the mental health of an individual, the plan is there. Your thoughts through the right devices can control what is happening in your brain or genes that might be affecting your mental health. In other words, when patients begin to feel the beginnings of an epileptic seizure or a deep depression, they could be trained to think in ways that begin the process of shutting down the chemical beginning of the problem. Their thoughts would turn on the light-activated cells that would pump out the seizure stopping chemicals. The same is true for depression. Light-activated cells in the brain would release chemicals to stop depression. Of course, we have seen many times that in human brains nothing may be that simple. Even with current anti-depressant drugs, face-to-face counseling is recommended to go alongside the pills we are taking. It is also well known that mental illnesses do not have one simple cause but can arise from a complex set of environmental and biological processes.

BrainGate2—Deep Brain Stimulation (DBS)

Current research is testing whether using Deep Brain Stimulation (DBS) as a treatment for Parkinson's disease can be effective in the control of symptoms. DBS involves inserting two electrodes deeply into the brain. An electrical pulse is then modulated in a specific region of the brain, assisting with Parkinson's disease, mood disorders, strokes, and pain. Rather than being a treatment of last resort, DBS is becoming more used by medical doctors immediately upon diagnosis of a problem. In Brain/Gate2 a baby aspirin-size sensor is implanted in the brain's motor area. The person thinks about moving a leg and the sensor picks that up and stimulates motor cells to fire. In time the act of walking can become smooth and coordinated and "thoughtless" at the same time. The same technology has been useful in sending thought signals from the brain to very successfully move an amputee's prosthetic arm.

Rewriting Bad Memories

Andrew Holmes, of the National Institute on Alcohol Abuse and Alcoholism, in a study on emotional reactions, showed that disrupting connections between the amygdala and the prefrontal cortex causes bad memories to persist. The amygdala becomes active during threatening situations, and when the threats are gone, the prefrontal cortex should become active and do away with the fearful memory. A properly working connection between the amygdala and prefrontal cortex is necessary for bad memories to be removed, and as such it may be an important direction in research on Post Traumatic Stress Disorder (PTSD). The amygdala and the prefrontal cortex are not the only players involved in the persistence of bad memories in people, however, since other brain connections are held in common by these two important brain areas. Strengthening that connection, perhaps, will allow normal treatments to remove painful memories in PTSD for example, as well as many mental problems involving anxiety and fear.

It has been found that when we manipulate living brain cells with high-frequency light bursts, a technique called opto-genetics, mice have recovered lost memories, a finding important in the search for treatments for Alzheimer's Disease. Again, it was connectivity that proved to be the key, only this time it was the regrowth of synaptic spines that improve the connectivity between neurons. Scientists at Japan's RIKEN-MIT Centre for Neural Circuit Genetics altered some neurons in the hippocampus of mice to restore memories of training that had been lost.

Recent research also points to what is called a Disappointment Circuit in the brain. Brain pathways involved in processing reward and punishment are also involved with depression. The goal has been to find better antidepressants that will affect only the activity of the disappointment circuit as it is called, and thus show more effectiveness against severe depression. With drug-resistant depression the use of Transcranial Magnetic Stimulation (TMS) has become the object of intense research interest. TMS involves the precise targeting of magnetic pulses to key areas of the brain. Such pulses will activate or depress key areas of the brain thought to be involved in depression. While it has been known that TMS can relieve depression, it was never known how this happens. Now the work suggests that it is the connectivity between different circuits in the brain that must be strengthened or weakened to help specific emotional problems like depression or disappointment.

The Ethical Realm We Are Now Entering

The best answers to ethical questions about the future brain technologies depend upon your particular view of human nature. Our moral philosophies may be informed by the neural sciences, but they cannot spring from them. A rigid reductionism is powerless to make any philosophical or theological suggestions about moral directions. To limit a science to just objective, experimental outcomes as suggestions for a technological future for human beings will not work if human beings are indeed more than just biological entities. Human beings desire so much more than food, and the world of the mind is so much more than mere survival. We so obviously have spiritual as well as physical natures. We are self-conscious, meaning-seeking, relation seeking, creative persons. If we believe those are the qualities of human beings, then our ethics must reflect that overall view of human beings and reality.

If we believe in personhood, remember what the new technologies cannot do. They cannot change sin nature, or bring happiness, meaning, and purpose to human lives. Tools like these can improve the work of the church just like they always have, *e.g.*, printing press, microphones, on-line learning, films, etc. But such developments will always be open to misuse as well. Do not fail to engage in scientific development simply because harm might come.

Leonardo da Vinci—A Person of Interest

Leonardo da Vinci was a genius who blended science and the arts, and kept the human person at the center of his work. Leonardo da Vinci was born on April 15, 1452, in Venice, Italy, and died on May 2, 1519, at the age of 67. He was an artist and intellectual with no equal during the Renaissance. His works include the enigmatic *Mona Lisa*, and the famous *The Last Supper* at the moment of the betrayal by Judas. Da Vinci was both a genius at painting and sculpture. He also had an amazing mind for architecture, military engineering, the technology of the future, and left thousands of unpublished pages of his sketches and ideas. He was hungry for all knowledge and in his journals he made a record of all that he was seeing, and left it all to us.

What is important here is that Leonardo, the great painter, sculptor, and forward-view scientific and human thinker, did not see so great a division between science and art. He acted as if all disciplines should be intertwined in life and study as they were in his own mind and life. He could see beauty and story behind all of his inventions and images in art and in the possibility of future inventions. He studied anatomy and dissected human and animal bodies in order to see human beings better. Science to da Vinci was a part of learning how to see, and not just how to control matter. He drew and left us a view of what he saw of bones and muscles and the head and heart, and the vascular system and a fetus in utero. Da Vinci filled dozens of notebooks with his drawings and thousands of pages of his observations of the world. His fine art and sculpture was all a part of his vision. He left us almost 2500 drawings of plant and animal life, human anatomy, and amazing looks ahead at what would be bicycles, helicopters, and flying machines, not usually to build them for use or to sell, but to amaze people at parties, and of course we are amazed so many years later as well. Da Vinci recently has been recognized for having the first experiments and written records demonstrating the laws of friction. He applied his understanding of these laws 200 years before anyone else when he described their usefulness in the workings of wheels and pulleys. Da Vinci's brain is not what is amazing here, but his mind is. It is interesting that da Vinci was actually more known for some of his robotics than for his artwork during his time. Only about two dozen paintings of his are known to exist because of his so many other and varied interests and knowledge.

Da Vinci was what we call a Renaissance man, who recognized no important differences between the sciences and the arts. C. P. Snow, a British scientist and novelist, author of the well-known *The Two Cultures* (1959), which still lingers in our discussions today, reminds us of what da Vinci showed us. Our western culture, Snow said, was split into two cultures, the sciences and the humanities, and that was a hindrance to our advances in solving personal, scientific, and world problems. We might do well to pay attention to this warning in the neurosciences as we struggle with the work of brain cells and Connectomes. We must realize that the outcome so closely connected to these biological materials is the self-consciousness that we all possess in full measure, so full, that it seems foolish to ignore ourselves in the midst of our study of ourselves. We are all familiar with da Vinci's circle with a man in it with outstretched arms to the edge of the circle. Art and science were together in this drawing called the "Vitruvian Man," who was a man superimposed over two forms, a circle and a square, with his arms and legs apart, everything in the art piece was one, not two, art and science together.

Concluding Thoughts

In the case of all these technologies we are the ones doing the making, not God. Maybe we ought to think a little more like God and not try to replace Him with the results of our own God-given abilities. And Christians cannot abandon the field simply because suddenly it sounds like a frightening immoral future. We can keep the excitement of discovery and control of our world under the lordship of God and the goals of personhood. I do not recommend for Christians to withdraw from the neurosciences because we think all things scientific or secular are bad, because they are not. But it does mean that Christians have to be well educated and enter into these professions at the cutting edge of technology and other fields such as the arts and the entertainment fields.

It is interesting that as the research is done with greater funding, a lot of the results that promise new and more technologies seem to be straight from the science fiction novels of the very recent past. We hear of mind-reading and dream-reading computers, humanoid robots, immersion entertainment, cyborg military humans, linking monkeys into a shared brain network, and more. The goals of these advances should be bound to the overall goal of improving human beings and human life with a larger theory of human nature in mind. It is nice to think that as our future and our interest becomes more controlled by the new technologies, that we would gravitate at the same time to the other culture of ideas, those that make us more personal, the humanities and the arts. But that does not seem to be the case thus far this early in the growth of the technologizing of the person and the culture. I remember "Star Trek: The Next Generation" when Captain Jean-Luc Picard—as he lived in a technological future world every day aboard the Starship Enterprise, with its computers, robots, and holodeck movie worlds—would find himself in his quarters at the end of the day, far from earth, sitting in a comfortable chair reading a leatherbound copy of Shakespeare's plays. We should not forget that lesson from an imagined man of the future, or from Leonardo da Vinci from the past.

And what will our future be? We are not going to avoid it, because it is already here. We can think with a belief in personhood about our future. We can guide and change the future in ways that fit persons. Are we to be an unhappy Frankenstein monster with a wicked brain, or a robot with no brain? Or as T. S. Eliot so well stated it about the future in this famous line "We are the hollow men, we are the stuffed men leaning together, headpiece stuffed with straw… for whom life ends not with a bang, but with a whimper" (from "The Hollow Men"). T. S. Eliot's thought seems to be more a real possibility today than ever before.

****

"If I only had a brain!"
—The Scarecrow, (*The Wizard of Oz*, MGM studios, 1939)

SOME BOOKS I THINK YOU MIGHT LIKE

***The Future of the Mind: The Scientific Quest to Understand, Enhance, and Empower the Mind*, by Michio Kaku.** Michio Kaku, a theoretical physicist, offers a look into the future of brain science and how these new technologies will change our daily lives. He writes about the latest advances in neuroscience on the cutting edge of recording memories, videotaping dreams, mind-to-mind communication, and more.

***The Future of the Brain: Essays by the World's Leading Neuroscientists*, ed. by Gary Marcus and Jeremy Freeman.** The stunning advances in brain science are highlighted with the implications they will have for medicine, psychiatry, and the psychology of consciousness.

CHAPTER 9

BOTTLED BRAINS

Malformed: Forgotten Brains of the Texas State Mental Hospital, a book by Adam Voorhes and Alex Hannaford, tells the stories of the 100 brains floating silently in large bottles of formaldehyde since the 1950s. Each brain is labeled with what disease or damage the brain had endured: Huntington's disease, Alzheimer's, gunshot wound, and so on. I recently had the occasion to visit the Indiana University Medical School museum where one room in the ancient building housed a large number of bottled brains and what killed them. There were cerebral hemorrhages, strokes, tumors, and more, and the damage to each brain was vivid and threatening to those of us with living brains. As I stared I again felt like Hamlet, this time looking not at a skull, but at a brain just recently alive. What was it in the cells of the brain that gave, not just life, but personal life? A lily is alive, but it is not you or me. I wondered all the while what my brain would look like some years from now in a jar of formaldehyde since a few years ago I had a burst brain aneurysm and emergency brain surgery. Would I see blackened, destroyed brain matter in my left frontal lobe? Would there be a hole I could peer into to see the clip my neurosurgeon put on an artery's rupture to save my life?

In neuroscientist's labs there are always dead human brains on the shelves or in classrooms somewhere, and more hidden in closets for undergraduate student labs. I suppose if they were out in plain sight that might be too thought-provoking, not necessarily about the question of will I survive into an afterlife – but the certainty of knowing that any one of us might someday be in a jar of liquid on a shelf in a closet, or on a lecture room table… or most likely in a decaying coffin under a pile of dirt somewhere.

A TITLE CAN SAY A LOT

Me, Myself, and Why

—Jennifer Oulette

THE END OF THE MATTER: BOTTLED BRAINS

> **"Why We Ask 'Why?' What is the ultimate truth about ourselves? . . . We are physical machinery—puppets that strut and talk and laugh and die as the hand of time pulls the strings beneath. But there is one elementary inescapable answer. We are that which asks the question."**
>
> **—Sir Arthur Eddington**

In 2015 an international team of astronomers determined that our universe is dying. By analyzing the decreasing energy output of 200,000 galaxies they found that many billions of years from now the universe will reach a point of heat death, not heat really, but a cold empty darkness. Of course, that is of little concern to those of us who will not be around on earth billions of years from now. All of us alive now will experience our own brain death before a hundred years have passed. That would be the time when Christians say that the soul departs from the body and just the shell of your former self is left. Dr. Duncan MacDougall in 1907 wondered about the weight of the soul, so he weighed some of his patients at death to see if it came with a loss of some weight. MacDougall knew when his patients with tuberculosis were close to the end, and so he moved them to an industrial scale and looked for a drop in weight when the time came. Sure enough, the soul seemed to weigh about twenty-one grams. Not surprisingly, very few scientists agreed with his findings.

If matter is all there is, then what is life all about for us? What is the meaning, purpose and significance of all our great endeavors that stand head and shoulders above the tooth and claw of the animal world, and seemingly above the silent universe around us? We develop a voice to sing and speak and then we are told that there is no one to listen to us. We have ears and yet there are no voices in the fading candle of the universe to speak to us. We are "All dressed up for a prom, and only a barn dance to go to."

In the beginning a materialistic and deterministic philosophy killed the soul, and then those who held that worldview realized that without the possibility of surviving the death of their brains, they, too, would be gone. So they steeled themselves behind their resolve to not worry and stay in meaningful lives by contributing books of information for the future; or they saw their function as contributing their own DNA and other assorted molecules and atoms to the memory banks of the universe; or they reserved rooms in a super computer's memory banks as the depository of their minds for the future.

To study the brain we must realize that, whatever our views on the soul—the immaterial essence of a human being—when our brain dies, our bodies must die to the world here. Our purpose and meaning is related to the world after this one, Christian theists argue, and the activity of our brains and ourselves is a look in an upward direction and not downward into the grave. But at the death of the body and the brain, our lives here are suddenly all over. So many cell divisions, so many cycles around the sun, so many hours to clock in, so many alarm clocks to set, and so much of real life it seems is gone forever. However, so much of the value of life, what not to miss in life, seems to be present here in the searching and longing for something beyond the world here. What are those things that we long for in this life besides physical survival in the form of safety, food, and sex? The family, the love, the sunsets, beauty and truth, love and justice, courage and heroism. We become filmmakers, artists, poets, authors, and neuroscientists. Where is that which we pursued in those brain networks and how did we miss it?

Or are we just freaks of development on the level of an eight-legged cow or a two-headed snake? We write of our longings, of our real hungers, in books to tell it to others, in some hope that it will not be missed. We transfer our mind's awe and fears and hopes on to canvas or musical notes or screenplays on to paper, or on printed or digital pages of Pulitzer prize-winning novels, or in our own leather-bound private diaries. Though we may die, indeed our books and our accomplishments do continue our great thoughts and longings, where they sit in libraries where you can read the thoughts of those from the past. What is it in us that longs for more than what is here? What do all our collected thoughts in the world's greatest libraries really teach us besides the ideas of math and culture and bridge building and particle physics and how Monarch butterflies manage to fly all those thousands of miles every year?

The Long Room of Books

The books scholars write are often touted as if the author continued life after the death of their bodies and brains. We should not mind death, it is said, because our books contain our best thoughts. But most books, especially the bad books, do die when they go out of print in about three years. In the words of John Ruskin, there are only two kinds of books: the books of the hour, and the books of all time. The books of the hour are those that are standing on forgotten and ignored book shelves in small, dusty, old libraries, but the books of all time are the Great Books that line the shelves in the most beautiful libraries in the world.

You can Google the twenty-five most beautiful libraries in the world, and usually ranking at about number three is the Long Room at Trinity College in Dublin, Ireland. The Long Room is beautiful, and it creates in tourists the feeling of awe at what human minds have achieved. That beautiful library with its row upon row of old books says that something else is going on in the human mind besides our impressive intellect. In print are our longings and hopes and dreams, and yes, our desire for something larger and more eternal than ourselves. The library was finished in 1732 and it is in the Old Library on Trinity College's campus. It houses some of the oldest books at Trinity College. The Long Room is so named because it is the longest single-room library in Europe at sixty-five meters.

The Long Room is a working library with supervised reading rooms at each end. What you see and feel as you enter the Long Room is that the human past is still with us and that its knowledge is important. The search for meaning and purpose has always been important and that search continues on in us who remain alive. This library is exactly as its title suggests, a long room that you walk down the center with books on both sides, two stories of books tall. There are 200,000 volumes, and the oldest volumes are older than the United States. Every author is dead, but their personhood has passed this world of learning on to us. The library is open to scholars who wish to study some old manuscript. In the long room, however, you would not be walking around the books, pulling those that looked interesting, off the shelves, but you might be hoisted, high in the air on an open little platform with a desk, and kept up in the air until you were finished. Try to run away with a first edition of Darwin's *On the Origin of Species*. I dare you!

Human knowledge is displayed out before us as we walk down the center of the isle in the Long Room, all 200,000 volumes, many of them the "Books of all Time." Their mere presence creates in us a feeling of open-mouthed awe. We do not use that word "awesome" appropriately any more. We say everything is awesome from apple sauce to a new song. But awesome should mean something that fills us with awe, or we are awestruck. The awesome feeling you get when walking through the Long Room is, here is a lot of knowledge, yes, but we also see human desire to go beyond this material realm. Yes, you can get most of these manuscripts that you would want to read online, on your iPad, or on your Kindle. Knowledge for human beings is collected, passed from scholars to scholars. We are communicating beings and we do more than shout "danger" or "food" or "sex." We have a solid timeline of our progress and the library represents that. In our little three-pound brains we have access to almost all of the knowledge of mankind for the last 6,000 years and even the revelation from God himself. Minds of men reach across the miles and the years directly into our minds and we do not have to rethink every important thought anew all over again for ourselves. Amazing! Thus we are able to build this massive civilization, which is recorded in this Long Room, and be in awe at the three-pound brain that did it all. Brains long dead continue to live and talk to us though their owners did die.

Aging and Death of the Brain

Since aging and death are more common to all of us than even taxes, you would think we would ponder about that subject more often. But we do not. We seem to put most of our efforts into improving our health and lengthening our lives. There is nothing wrong with being healthy and living long, youthful lives, but the end game in life is not about living forever here. At the cellular level we know that our bodily cells manage about fifty-plus cell divisions before they begin to multiply mistakes. The brain's neurons for the most part do not replicate, although there are new cells being produced in certain areas of the brain such as the hippocampus. On a tinier level we know our telomeres, those repeated DNA sequences at the ends of our chromosomes, get smaller with each cell division until they are just about gone at fifty subdivisions. Then cell divisions cease and cells die. The telomeres' function may be to keep the chromosomes and our genetic code operating at peak accuracy levels. The plastic tips on the ends of your shoelaces serve a similar function in keeping them from fraying and you from angry efforts trying to push the frayed, wild shoestring through the eye holes in your shoes. Some aging research centers on the idea of preventing the telomeres from shortening and thus prolonging your life. But like most things with the complicated human biological being, such research has not had an easy path.

On the more practical side of aging there is obviously some decline in memory in the elderly. At first it was thought these difficulties to be related to the loss of gray matter, or unmyelinated neurons in the elderly person's brain. Though the brain matter decline is there, it has turned out that it is the loss of cells in the hippocampus that correlates with mild impairment in memory that we are used to seeing in aging adults. However, most cognitive abilities show little change as the adult ages until reaching much older ages, generally not until their eighties and older. It would be a mistake to think that mental decline is inevitable with age, since mental ability does not seem to correlate exactly with brain matter loss. The elderly may have fewer neurons and a bit less memory, but that does not affect the aged terribly if they stay mentally active.

However, the elderly remaining mentally healthy cannot be said to be the case with Alzheimer's disease, a brain disease largely in the elderly. It would be an error to think that Alzheimer's disease is the result of normal wear and tear on the aging brain. Alzheimer's disease is rightly called a disease, and it negatively affects those areas of the brain that are important for who you are as a thinking, willing, feeling, self-aware person. Over 4 million Americans suffer from Alzheimer's disease and that number has been on the rise for years, since an increasing lifespan in the elderly has come with an increasing percentage of Alzheimer's in that population.

Concerning Alzheimer's disease we are used to hearing by now about the buildup of amyloid plaques in the limbic system and cortex, the presence of neuro fibrillary tangles in the brain's nerve cells, and the reduced production of the important transmitter ACh. The Alzheimer's patient's brain shows increasing cortical atrophy in frontal, temporal, and parietal lobes of the brain, which I suggested in chapter two show most of the attributes that relate to our personhood when interconnected with other brain areas. The end result of Alzheimer's disease is personal death and then physical death. The patients have been called the walking dead for this reason. Perhaps we all have seen a person we know who has Alzheimer's, who gradually becomes the individual standing blank eyed next to a spouse or friend. We all remember the Hollywood actor and president Ronald Regan in this way, and more recently the singer Glen Campbell. Unless people have a view of life that transcends this world and death, then aging and death become our major enemies. We want to lengthen the years we live, improve our health until the final end, and for some the goal is to eventually achieve immortality by transferring our consciousness and memories into a super-computer.

Progeria Kids—The Rapidly Aging Child

If our normal journey to old age, disease, and death is not thought-provoking enough there is the condition of progeria, the lightning-quick aging disease in children. The average person does not think about death until he or she is faced with it. But if we all went from birth to very old in 10–12 years, we would think about it all the time! Progeria is the very rare disease of rapid childhood aging. We normally see a progeria child in the news taking a trip to Disney World as a token to a life never lived. For that child who looks to be in his eighties, his body already has heart disease, a bald head, tight skin and wrinkles, lost teeth, and brittle bones. For these children there will be no high school athletics or homecoming dances. There are usually no college days or dates or marriage. There is only the rapid decline and advance toward death.

If this was a planet of progeria victims, then rapid aging would be a normal state of affairs. Everyone is born and lives perhaps a few years into puberty and dies of old age. If this was a planet, on the other hand, of life to 900 years, our eighty or ninety aging years would be seen as a great tragedy. It is interesting that our minds do not feel as old as our bodies. Outside we have an 80-year-old body, with no more football or modeling career for us. But inside we feel like a normal 35-year-old. We hold on the handrail for steps and have aching backs every time we move. The mind seems so much more capable than our bodies for living and thinking, giving out eventually when the brain's decline destroys those areas which seem to be in partnership with our mental selves.

There is another disease in children that is the opposite of progeria. This is an extremely rare condition with only a few cases worldwide, where the child is prevented from aging. Imagine a child, who still has the soft skin of a newborn and though she is eight, she weighs only eleven pounds, and wears three- to six-month-old size clothing. A 29-year-old man from Florida has the body of a 10-year-old, and a 31-year-old woman from Brazil looks like she is two years old. Research on this condition still has not yielded any answers as to why the developmental process has slowed to almost stopping. Medical researcher, Richard F. Walker, at All Children's Hospital in St. Petersburg, Florida, is researching the condition, in which his patients age at about twenty percent the normal rate.

The genetic search in the battle against this disease includes looking for information that would retard aging in all of us, the search for longer, healthier life and someday, perhaps, immortality. This disease is also known as Benjamin Button disease after the Brad Pitt movie, based on a novel by F. Scott Fitzgerald, which is about an old man gradually getting younger. This disease has been described as developmental inertia. Does this mean that there is an off switch for development, and what would we do with it if we found it? Turn off the process at age twenty-five and let it tread water to keep you at a constant age and body? That should satisfy us, right?

But getting younger seems no better than getting older as seen in all of these diseased children. Who wants to look like an infant or a child when you could be older mentally? There was something of this theme in the book *Something Wicked This Way Comes* (1962), a dark fantasy novel by Ray Bradbury, and made into a Disney movie with the same name. There was an evil nearby and it centered on a merry-go-round on which adults would ride in order grow younger. Good idea it seems, but it had tragic results. It seems as if there is more to what we desire in life than just getting younger in an aging world. I am not criticizing the use of Botox, the cosmetic to remove aging wrinkles on the face. What is better, perhaps, in a world in which we must die, is to become content with who we are as persons, in healthy aging bodies, but persons nonetheless, in bodies that will not make it out of this life alive, even with Botox.

We can fulfill our purpose in life whether we live to thirty or ninety. Doing research in medicine to help ourselves make it to 102 with good health is acceptable and beneficial to some of us, but what are we doing at the childhood end of the human race for infant mortality? What about the poor of the world who will not be able to afford the age-lengthening medications or surgeries? We will soon have the potential to change the human race in many ways, and when that day comes, we need to have asked what value are we seeking in our extension of the life and potential of the person.

Freezing Bodies and Brains

One thought to holding off the inevitable death we must face is to freeze the head and brain until the future age when the ills of the body can be healed by future medical advances. Currently when death has been declared because all activity in the cortex has ceased, a dead body may be frozen as quickly as possible until the organs of the body can be harvested for waiting patients. Freezing of course also happens at cryogenic centers that can even intercept a patient at the moment of death and then begin the process of keeping the blood flowing, restoring oxygen to the lungs, and eventually replacing the blood in the body with a medical-grade antifreeze. The body is then stored in liquid nitrogen at -195 degrees C for hundreds of years if needed, until science can cure the disease that killed the patient. Then it will be the time to awaken the dead. The world's largest cryogenics facility is Alcor Life Extension Foundation in Scottsdale, Arizona. The biggest uncertainty, however, with this plan is whether the process can really bring the human brain and mind back to life to function normally. The cost at Alcor Life Extension is $220,000, payable upon death.

The Immortal Mind

To transfer the human mind into the circuitry of a computer, of course, assumes that there is nothing else to us but mechanical memory circuits, and that you are your memories. If we can transfer all your memories into a computer, then you and all your abilities to think will still be there. Research has shown that we can transfer simple memories in mice into a supercomputer, and even then move them back into the mouse's brain or another mouse at a later time. This does not mean that we can transfer a knowledge of the Russian language into your hippocampus or transfer your life history of memories into a computer bank. Of course, to do such a transfer we would have to map your brain right down to the basic level of activity, every memory stored, every thought produced at this moment, perhaps even down to the atomic level in and around synaptic clefts. This would be the real Connectome of you. This Connectome of stored information would become a computer model of you that would work like a virtual brain. At least that is the long-range plan.

Right now, with the hard problem still nipping at our heels at every step, we do not even know what conscious experience is, or where it resides if any "where," let alone know if it would exist in the digital switches of a computer. Ignoring the hard problem as we plan the future digital existence of you is just a head-in-the-sand approach to the topic. But the "science of the gaps"—give science a chance—keeps assuring us that we will know more about that in the future because, after all, the brain is all there is to you. And, we would certainly know that you were now in the computer because you told us you were. I guess we would have to assume you were not a philosophical zombie—no mind, but speaking like it did have your mind. The Turing test would say that such a positive report from "you" in the computer is good enough to say you were mentally conscious. And then you could live forever, an immortal human being, as memories in computer software. That is your future, our future, a second evolution, the first one biological, the second one through digital transfer. The first one slow, millions of years. The second one rapid, under our own control. Death would not be our enemy, but will we conquer it with the machine mind?

Can You Transfer Your Consciousness into Another Body?

Mapping your brain Connectome and collecting all your memories and learned and genetic programs that shaped your brain at one moment, is not going to be simple, and may indeed be impossible. The best Connectome so far of the mouse brain is one done by Harvard neurobiologist Jeff Lichtman, of a piece of mouse brain smaller than a grain of sand, which is several billionths of the brain, and took six years and over twenty researchers to complete.[1] That brief Connectome was a dense 3D map of the mouse's neocortex.

The map takes into account every branch of every neuron and supporting cells included, and the connecting locations of 1700 synapses. The next project for his lab is to do a map of an entire cubic millimeter of the mouse brain. We have a long way to go to getting a Connectome map of the human brain, and then it would only be the neural connections, which hopefully includes your memories. If some new theories about subatomic storage of memories comes along in the meantime, the Connectome of you may never be done. And the true Connectome of your brain will have to take into account the brain activity away from the synaptic connections, down even to the atomic level, and the myriad of highways and alleyways that are chemical in nature. A Connectome map is a good, useful start on understanding your brain, but it is doubtful that we could ever call it you, all of you.

Neuroscientists' Thoughts on Death

What follows now are some typical quotes from well known neuroscientists and neurophilosophers, who are leading thinkers and researchers in the field. They are well respected. I like the work they do, even if I disagree with some of the interpretations of their research. Their comments, just a short quote from each, are from Dr. Susan Blackmore's excellent interviews with some of these experts. What I want to show is that the field as a whole believes in a materialistic, deterministic model of the human person. And if you espouse such a materialistic view, it leaves you little room to say anything to the question, "What happens when you die?" They feel that their scientific research has shown them, almost certainly, that with the death of the brain there is the end of your existence.

David Chalmers: an Australian philosopher interested in consciousness studies. "I don't know for sure. But I'm inclined to think that my consciousness ceases to exist.... After death, my brain will disintegrate, so my consciousness will disintegrate too. ... Then again, no one understands consciousness, so I could be completely wrong. That would be nice!"[2]

Patricia Churchland: a philosopher working at the boundaries of mind and cognitive neuroscience. "All the evidence shows that the brain is necessary for functions associated with consciousness. I am not sure how consciousness could survive the death of the brain if it needs neurons to sustain it. At a personal level, I should say that I feel more settled about death and dying having understood that it is the end, than I would if I were trying to nourish an unrealistic hope in some kind of heaven."[3]

Paul Churchland: a philosopher, and husband of Patricia Churchland. "Consciousness is just one particularly sophisticated dimension of biological life. When my biological life ends, so does my consciousness. I am more than content with this. The prospect of being conscious for an unending eternity is quite frankly appalling. When my time comes, let me sleep."[4]

Francis Crick: Nobel Prize winner as co-discoverer of the architecture of DNA. His later years he spent in collaboration with Christof Koch on their search for the neural correlates of visual consciousness. "Personally I believe that it is highly unlikely that there is consciousness after death, but that, after all, is what we are attempting to prove, in as far as one can prove anything scientifically."[5]

Vilayanur Ramachandran: originally from India, with wide-ranging interests in the brain and cognition. "It's enobling, rather than diminishing. It's only when you start thinking that you are some aloof thing which is in charge of everything, that you become scared of dying, because you say, 'Oh, my God, when I'm dead, I'm not around anymore.' But if you think you're part of the ebb and flow of the cosmos, and there's no separate little soul, inspecting the world, that's going to be extinguished—then it's enobling. You're a part of this grand scheme of things."[6]

Carl Sagan and C. S. Lewis—Persons of Interest

These are two famous scholars, who have used nonfiction and science fiction to speak about their views of the universe. They developed their careers as they searched for meaning in life and answers to the ultimate questions in life. While they both wrote extensively in nonfiction, both used science fiction to give their answers to why the universe exists and the meaning and purpose in this life. Lewis' answer to the meaning-in-life quest for human beings was well illustrated in his book *That Hideous Strength*, the third of his science fiction trilogy. Carl Sagan showed us his struggles with the same question in his book, *Contact*, also made into a movie of the same title.

Carl Sagan and His Universe

The Library of Congress has all the papers of Carl Sagan, the astronomer, because he was so interesting as well as famous. His universe was both beautiful and powerful, and also dark and empty. He placed human beings, and every great accomplishment on our planet, as a bit of nothing in the terrifying beauty of it all. Why would we be the only ones awake in this massive and beautiful universe? Was it not reasonable, he thought, to expect there to be other civilizations out in the universe who might make contact with us and explain all the mysteries of life to us?

Carl Sagan was a brilliant communicator, and he brought millions of American homes a vision of the universe with his hit television program, *Cosmos*. He began every stunning episode with the line, "... the universe is all there is, or ever was, or will be." He did not believe in a higher being who ruled the universe with the name of God. But he always seemed to search for the meaning in the natural order, and later he searched for undiscovered life out there who could give us their wisdom on who we were and why we were here. You can see that in his search for extraterrestrial intelligence, the SETI project featured in the movie. And his book and movie *Contact,* which starred Jodi Foster and Matthew McConaughey, as the empirical scientist and the nation's pastor respectively, whose lives revolved around the discovery of life out there, and dealing with faith in the process.

Contact showed Sagan's constant search in this universe that he hoped would bring some meaning to it all. In *Contact* he showed us the comfort he received knowing that there were nice, powerful beings out there watching over us. Their presence brought enough comfort to Jodie Foster's character in the movie that she became more content with life and less distracted by what had previously annoyed her as she listened with the radio telescope for the sounds of alien intelligence. Sagan's view was that science was not only compatible with spirituality, it was in a way the source of spirituality.

C. S. Lewis and His Universe

C. S. Lewis was called a most reluctant Christian convert, dragged kicking and screaming into the Kingdom of Heaven. He was a brilliant atheist, who later became a great apologist for the Christian faith, in both his nonfiction and fiction. He wrote seven children's books that spoke deeply about the reality of the Christian worldview, and three science fiction books that took us out of this world so that we might see this world more clearly. His children's books were also for adults, who could still see with a child's eyes of mystery and wonder.

Lewis was born, raised, and became a scholar at Cambridge and Oxford. *Surprised by Joy* was his spiritual autobiography, in which he described the pull that he felt toward the ultimate things in the universe of the mind—toward love, courage, heroism, truth, and more. And he could not explain why there would be deep desires in us for anything other than bananas and grass. Either, he thought, the universe was bigger than he was being told, or human beings were only freakish products of chance evolution. Lewis could not come to the conclusion that we were freakish animals, since human beings created with such beauty and wisdom.

C. S. Lewis is one of the most-read Christian scholars today in both his fiction and nonfiction. His *Out of the Silent Planet* begins the travels of the philologist named Ransom, an expert at languages, to Mars, then Venus, and later the trilogy culminates back on earth. Lewis gave us a clear picture of the universe, where he found the truth about the God who was out there, and angels, who were beings of love and power. Through the lead character Ransom, Lewis showed the reader a high human purpose and calling, which gave meaning to human life and meaning in times of suffering. He also showed the "bent" nature of man, as he described the sin nature of human beings. What Lewis did say so clearly was to not ignore the deep longings of the human heart. The deepest desires of the heart, he felt, argue that we were made for a larger world.

Both Carl Sagan and C. S. Lewis are good examples of the deeper longings of the human heart. Human beings are not just searching after grass and hamburgers, but searching instead for the ultimate things in life, the longing for the perfect, the courageous, and the true. These longings are seen so clearly in our science and our arts. It is interesting that the longings we feel inside are just the opposite of what we see in a world gone bad, all the things we do not like. My longings replace love for hate, understanding for ignorance, heroism for cowardice, truth for deception, freedom for slavery, and humility for arrogance. Lewis spoke clearly of finding what he longed for. Sagan was still searching at the end.

Concluding Thoughts

Death should make us stop and think of our great longing for life, for real life, not like an antelope struggling to get away from the jaws of a predator, but like a starving man following a smell of food. We, the possessors of this wonderful brain, long for something else beyond the food and safety of our daily existence, and hope for something else beyond this aging body. We should prepare for death as a doorway through which all must walk, rather than ignoring or running away from the certainty of the end of the body in this life. There is no reason not to try to be the best physical specimen that we can be. But it is better to be a better person than a better body. Whatever the relationship between the spirit and matter of our consciousness is, the Bible chooses to describe that mysterious separation at death as the spirit unnaturally separated from the body for a time. But later, when death is undone, and with the resurrection of the body and the wholeness of our nature is formed again, then we are human beings again, incarnated as conscious persons forever.

Christian theology teaches that life is an adventure, but it is not the last adventure, and those persons who trust in Christ will be made new and be able to enter the universe as God's children. They will enjoy and use His gifts to us. It will be time without time and the longings of each person will be filled in each moment, particularly love. The idea that there is no life after life is just an assumption that has swelled way beyond its merits as another faith to guide our lives. The immaterial essence, the self-conscious experience up there in our heads, which is our immaterial consciousness and self-consciousness, is evidence of the true nature of the universe. To be willing to believe will open up many more doors and visions to see and understand what is going on in that three-pound universe up there on our shoulders.

****

"You have made them a little lower than the angels
And crowned them with glory and honor.
You made them rulers over the works of your hands."
—Psalm 8: 5–6

SOME BOOKS I THINK YOU MIGHT LIKE

***How We Die: Reflections on Life's Final Chapter*, by Sherwin Nuland.** This bestseller and National Book Award winner, has chapters on the most common ways we human beings die, e.g., heart disease and cancer. There is even a chapter on murder, which is not always a bad way to go according to Dr. Nuland. With his excellent writing he makes the point that we Americans really do not know how to die or to approach the inevitable.

***Gratitude*, by Oliver Sacks.** This short little book of essays by Oliver Sacks was written when he received the medical news that he was dying of cancer. He records his thoughts about his own life, including his homosexuality, and all along gives his view that living a good life was achieving a peace within oneself. He felt we should try to reach the Sabbath rest of one's life with a gratitude for life itself.

CHAPTER 10

THE ARTIST'S BRAIN

Dr. Gregg Dunn is known as the "neuroscientist-painter," first because he has a PhD in neuroscience from the University of Pennsylvania, and then because he paints neurons, glial cells, and brain structures true to the science of the brain. His beautiful work is an accurate scientific look at the human brain with an ancient East Asian brush-painting technique called "sumi-e." Those Asian paintings are intended to show the artist's mental and emotional state, and Dunn's paintings do show the brush strokes of his mind's thoughts about his brain. Dunn says that for him the act of painting is an excellent environment in which to grow spiritually. What is important to Dunn is that he thinks he can bring emotions and beauty to a rigidly empirical field like neuroscience. His work is intended to transform the way ordinary people view the brain. He is able to create feelings in viewers about the importance of the brain to our lives as persons. He hopes his art will inspire a new generation of artists and neuroscientists.

Greg Dunn usually starts with tiny drops of gold or black paint dripped onto waterproof paper. Then with strong puffs of air through a straw he turns the blobs into finger-like extensions stretching outward that quickly become dendrites and axons and glia. In this way he imitates the actual tree-limb-like complexity of neurons and their connections. Dunn also has beautiful designs of the brain in gold leaf showing the brain Connectome and the artist's thoughts all at once as he paints with gold and puffs of air. Greg Dunn seems better with paint to fuse brain facts and art than experimental articles or fMRI scans. The human brain is the most complex physical structure in the universe. According to the neuroscientist painter it may be the most beautiful.

A TITLE CAN SAY A LOT

From Cells to Souls-and Beyond

—MALCOLM JEEVES

THOUGHTS ABOUT PERSONS AND BRAINS: YOU WERE NEVER OUT OF MY MIND

"What is man that you are mindful of him?"
—Psalm 8:4

The human brain has a face, and it is attached to the skull an inch or so from the mysterious inner workings of the brain. The human face is unique in many ways that indicate the personal nature of both the brain and the face. Face to face we meet, we talk. We face our problems, we face the future, we face our own limitations. The fusiform gyrus, also called the fusiform face area (FFA), is in the middle of the brain, the bottom area of the back of the brain, on both sides of the temporal cortex, and it is also known as the face-detection area. Damage to the fusiform gyrus can produce prosopagnosia, the inability to see a face (prosopon: face; agnosia: to not know). So important is your face that if you lose your face through a terrible accident or act of war, we are willing to try face transplants that are just short of reasonably okay. But they will get better.

The face is able to reveal what is going on in the brain with its several dozen, paper-thin, subcutaneous muscles attached to the underside of the skin of the face. They move skin, not bone. Thus, both our emotions and motivations of our inside world are present for the outside world to see. The brain makes two-way connections to the face with your feelings and thoughts moving facial muscles and your facial muscles sending signals to the brain. Thus, not only do your brain's inner thoughts and emotions change your facial expression, but in the reverse, to a lesser extent, your facial muscles of smiling or frowning can influence what you are feeling.

People with prosopagnosia do not see the correct world above the neck, sometimes just a floating cloud of gray or floating face parts. Some damaged by prosopagnosia will see the same head and face on everyone's neck, or see a simple caricature on every face, like a cartoon look-alike of the person. Others see a familiar face on a person but have the emotions for some other character such as Adolf Hitler. At that time your experience becomes the invasion of the body snatchers and my mother has been taken over by the aliens.

Faces are more important to human beings than we realize, because people and relationships are important to us. With that in mind, the fusiform face area does not work alone. This facial detection area does not just help us see faces, but it also develops the emotional input that we feel for that face, as well as the biography of that face. In other words the facial detection area contributes to my view of you as a whole person. I am a person with a biographical history and you can feel emotions when you see me. Inputs from the brain's limbic system, the emotional motivational center, are connected to the fusiform gyrus. The person we see also means something to us as the parietal lobe and frontal lobe connections are made with this very important piece of brain. Therefore, the fusiform gyrus constructs a face, we recognize it as a face as opposed to a car or something else, and we know the person's background and whether we like or dislike them.

Oliver Sacks, the famous neurologist with such a special human approach to brain disorders, reported several years before his death that he had prosopagnosia, and he had for most of the years of his life been unable to remember faces he had seen. He was perhaps best known for his book, *The Man Who Mistook His Wife for a Hat*, with a chapter of that title about a man with severe face blindness.

Given the large brain areas involved in person recognition, we might properly say that the human brain is more prepared to eat dinner with someone than to search for someone or something to eat for dinner. Persons and relationships are meaningful for us. Meaning means connected. If I gave my students a "Q" for a grade in class, they would want to know what it meant. I could tell them it meant "quick" and they were the best students I had. I could give them three lines for grades which mean nothing. But if the three lines are connected to make an F, their reactions would be much different than if they were connected to make an A. To ask if life has meaning is to ask is my life connected to anything important beyond just eating and sleeping. How is my existence in the now connected to anything more permanent? Do I have any relationship with the depth of persons, or just to bananas and breeding in the now?

In this chapter I will be giving some concluding thoughts on the mind/brain. It is interesting the way I just said that. "I will give thoughts on my mind/brain." It's Pooh bear all over again looking at the Pooh doll. How can Pooh ever really understand Pooh with only a Pooh-like brain? I guess I must again say I am in agreement with the closest of connections or incarnation of spirit to flesh in the mind/brain issue, but that does not mean that I think brain equals mind. There is a self, a person, who is peeking out behind all the data gathered. It is in the incredible abilities shown by human beings that should make us reconsider the nature of the person as more than just matter. It is in the incredible pursuits of meaning by human beings that indicate something else is going on up there other than the traditional efforts to munch on food and date and stay away from dangerous wolves.

And then there is the unexplained self-awareness that is viewing the awareness part of consciousness. And we all believe that self-awareness is special indeed. Look at Penfield's patient's reports as they are on the operating table, knowing themselves, and yet they also experience their own conscious experience, as opposed to becoming the experience as we do in our dreams, which have an absence of self-awareness. What does all this mean? I, myself, am here, both tightly bound to flesh and yet separated in ways hard to describe, a prisoner of the flesh at times, but master of my ship at other times. I am that strange creature on the earth, a person, on a seemingly impersonal planet, looking for personhood in the universe to explain me.

Important Ideas To Remember in This Book

Assumptions are important.

This is because they have consequences, and they do affect science. Assumptions or presuppositions or any types of beliefs certainly affect the subjects that you choose to study, whether you invest your efforts in the study of the brain, in phrenology, in scanning techniques, in the Connectome project, or in the interactions of brain areas. Assumptions influence how you design experiments, and how you interpret and choose to apply your data and its discoveries. Materialism (all is matter/energy), radical empiricism (empiricism is the only way to really know) are not well supported just by saying no one educated believes in God anymore. They are not reliable assumptions, even for science. They are supported by circular reasoning, *i.e.*, all is matter, therefore I will only use empiricism in my studies of the brain, and guess what, I have discovered people are only material beings. Radical empiricism, with its circular reasoning, has done more to bias neuroscience away from the belief in personhood than any other assumption. And remember, perhaps the major support for the personhood of humankind is that we are carrying it up in our skulls. The belief in personhood is what all of us feel up there in our heads. And then we see the outworking of that belief in the accomplishments and motivations of humanity as a whole.

Subjective experience is important.

Subjectivity exists just above your shoulders in your head, in your very brain, and even in sensory contributions from the body itself. You cannot ignore the reality of your own feelings of thought, emotions, and will. You can say that it is possible that everyone else on the planet is a zombie, without thought or feeling. All other people are just robots. But you should not deny your own experience and still be taken seriously in neuroscience. To claim that your experience of consciousness and self-consciousness is just an illusion leaves unexplained the illusion of experience itself. Why am I experiencing this illusion? Therefore, the conscious person has to be a part of our study in the journey to explain the human brain. That approach would change some methods and open up some new interpretations of data from the very beginning of the work we do. Subjectivity does not pollute the objective fields of neuroscience since the subject matter of ourselves itself is filled with subjective experience. I am really here.

The hard problem is important.

If we are willing to work seriously with the hard problem, it is going to change our very concept of the material universe because there is a key piece of the universe that is partly non-physical and subjective. And the scientific method will have to be altered to not just study the biology of human beings, but the nature of the universe itself. The human person is the evidence that materialistic monism cannot be held any longer when dealing with the human person or the nature of the universe. The human person is a picture of what is true in the universe. We need to make clear what science can and cannot do. The role of neuroscience in a search for a complete understanding of ourselves needs to begin with the idea that we do not know everything about the most complex physical structure in the universe, the human brain.

The unity of human experience and brain function is important.
The Connectome is important, but it is not the only avenue to pursue, and in fact the Connectome is bigger than just active neural highways. There are the constant formations and reformations of the minor axon highways, the synaptic connections and the chemicals making those connections. Any Connectome that is going to be helpful has to include gray matter, the billions of neurons with no myelinated axons, the majority of neurons in the brain that exist in the cerebral cortex and the cerebellar cortex. And then there are the small and large chemical-modulating highways that play an even more important role than the Connectome per se in the human personality. We do at times look more like a poem or song when you consider the rhythms of the large-volume transmitters, than the maps of the Connectome if it can ever be done completely. The human brain is the most complicated thing in the universe, so why would we think simple brain parts or simple explanations would describe it? As in this case, Occam's razor, the simplest explanations up here in my brain, are probably not the best explanations.

Top down thinking is important and even necessary.
We all have theories and assumptions, like it or not, which come from above. Not from the pope, but from our life experiences, from our views negative or positive on religion and God, and from our particular subject areas of interest and expertise. To argue that in science we can remain unbiased in research because we have control groups, and we have refereed journals that help keep science honest and correct, does not remove the effects of prior assumptions, which are stubborn in neuroscience and largely unexamined. If we do not examine our prior assumptions about the nature of the universe and the existence or non-existence of God, then we are a prisoner of the effects of those assumptions, even in neuroscience. An assumption like radical empiricism is not the best path to knowledge in an extremely difficult and complicated subject like the human brain. You cannot self-correct away from the negative results of wrong assumptions with just control groups and refereed journal articles, if the truth about the brain does include the subjective person in addition to the objective reality of matter.

Top-down does not mean the pope screaming at Galileo and neuroscience. It means there is something in different levels of the subject matter that have a bearing, and not necessarily a horrible bias, on what you are studying. You do not have a full understanding of neuroscience without considering the person, who is the subject matter of these brain investigations. To just ask a person to think of the number five while she lays in an fMRI is not the beginning of understanding the brain's involvement in discovering calculus, nor the motives of the great Isaac Newton himself. So much evidence points to some personal top-down theories of the person that would be helpful in this regard. Calculus is more than machine progress. So, also, is dancing, and poetry writing, and so are the Hubble pictures and your reaction to them, and so is feeling sorry for progeria kids, and your worries about death. I am not just claiming that science cannot explain these things in time, I am saying that there is much evidence in the human brain that says something else is going on in the brain other than neural activity, and the field of neuroscience will be partially paralyzed in its pursuit of explaining the brain unless it considers the whole picture.

Personhood is important and the key to understanding the human brain.

We must say that we are the key subject matter in the universe. We are the key that will unlock everything we need to understand in the universe around us. That does not mean that we do not need all the fields of knowledge. Nor does it mean that we do not need an understanding of God and His relationship to us. We are a part of the matter of our brains and bodies, and we are also above the things around us and in us. There's a mixture of the sacred, the world around us, and the world in us—and that mixture is not a dualism that completely separates the humanities and the sciences in our studies and in our research. In fact, we are the central subject of all, the point at which all subjects combine because we are both in the universe and it is in us.

You cannot deny personhood simply because you have chosen methods that never look for it. As I said before, that is being guilty of circular reasoning. You believe there is no God or spiritual elements in the universe, and therefore, you use only radical empiricism to gather knowledge, and, therefore, you only discover matter. Blinding ourselves to anything but a material world means that there can be no meaning in life, and no purpose either, and we label that outcome the anthropological crisis. We are persons and neuroscience can be missing the truth about us in its strong pursuit of empiricism and reductionism, and bottom-up-only methods, and its antagonism to religion and theism, and its arrogance in thinking it has almost explained the whole ball of brain (promissory materialism). The brain is the most marvelous thing I know, and I know it because it is me, and it is you.

The poet agrees. The brain is wider than the sky, contemplates Emily Dickinson, in a poem of the same name, and yet I can hold the entire thing in my head.

The Brain is wider than the sky
For, put them side by side,
The one the other will include
With ease, and you beside.[1]

I am not trying to separate science and religion. I want to look at brain science with a better view than a simplistic view of materialism and determinism that excludes the mysterious matter of me. I was not trying to have all the answers or a theory on how the brain operates. The brain is like the ocean and we have just stepped off the shore and are wiggling our toes in the watery sand. I wanted us to see the wonder of the brain and the persons who possess this matter of mind. I wanted to suggest that Christianity is a thinking person's faith, and Christians can think very well in the sciences without being negative and condemning. With all the advances coming in neuroscience, we need more Christians in the sciences, not to guard against the advances, but to use their unique point of view to point the way in new and original thoughts and applications.

Marilynne Robinson—A Person of Interest

The human mind is amazing, and it is incredibly difficult to understand the nature of the genius of some people. Some individuals have thoughts and words that go far beyond our ordinary minds, which are impressive in themselves. All of our ordinary minds go well beyond the signaling of a chimp, but some human minds leave the very stuff of ordinary human life and move into a made-up world of people and words and imagination and then slip back to real life again for our understanding. This describes Marilynne Robinson, who is one of the most highly acclaimed writers of our time. Her novels make us see what we did not see before and understand what we saw but never deeply. With her novels she looks past the actions of people and sees for us the thoughts and feelings in real characters that neither they nor we understand. She is a writer who is searching for her own motives, following beauty or tragedy, feelings that are setting the tone for thought, seeing past the mist into something bigger than war or death or oneself. How does the human brain do that?

Marilynne Robinson is the famous Pulitzer Prize-winning novelist who recently made the news when she alone was visited by President Barack Obama in a stop-over in Des Moines, Iowa, on September 14, 2015, while he was doing presidential business. He said he just wanted to have a conversation with her about some of the broader cultural forces that shape our ideas. She is the author of *Housekeeping*, *Gilead* (the Pulitzer Prize winner), *Home*, and interestingly a book on the brain called *Absence of Mind: The Dispelling of Inwardness from the Modern Myth of the Self.*

She feels strongly that the older, truer truth about the human self has been discarded by the newer proclamations of a science that seeks to explain consciousness and self-identity and even religion with just its biological tools. With her book on the brain she asks the hard questions of neuroscience and its conclusions on the brain. Marilynne Robinson shows her superb arguments in her essays looking critically at science. She argues against the damaging assumptions of neuroscience, and their blindness to their own presuppositions that lead them to discover exactly what they believed anyway. She argues for the person, not losing us in the weight of objective terminology in modern neuroscience, as opposed to the reductionionist materialism of current neuroscience leaders.

T.S. Eliot—A Person of Interest

T.S. Eliot, who was born in 1888 and died in 1965, is another genius—with words, ideas, and the poetry of life. He was born in the United States, but became a British citizen in 1927. T. S. Eliot was one of the modern world's greatest poets, a brilliant atheist who became a brilliant Christian. He received the Nobel Prize in literature in 1948. He did not ask for, but he got, our attention with our modern world's issues and problems. He called to the world to look around, and really see, and think about what we see. Eliot is the poet of the searching *The Wasteland*, and the complex and beautiful *Four Quartets*, and the poetry book of cats, *Old Possum's Book of Practical Cats,* which became the words and themes of the musical "Cats." Listen as he talks to the world in the modern age.

> "The endless cycle of idea and action,
> Endless invention, endless experiment,
> Brings knowledge of motion, but not of stillness;
> Knowledge of speech, but not of silence;
> Knowledge of words, and ignorance of the Word.
> All our knowledge brings us nearer to our ignorance,
> All our ignorance brings us nearer to death,
> But nearness to death no nearer to GOD.
> Where is the Life we have lost in living?
> Where is the wisdom we have lost in knowledge?
> Where is the knowledge we have lost in information?
> The cycles of Heaven in twenty centuries
> Bring us farther from GOD and nearer to the Dust."
>
> —T. S. Eliot, from *The Rock* (1934)[2]

Children—Little Persons of Interest

I Never Saw Another Butterfly is a beautiful book of little children's drawings and poems from the Terezin Concentration Camp from 1942-1944. Yes, I said children and a World War II German concentration camp. When Jewish parents were taken away to concentration camps and their eventual slaughter, their children were taken to Terezin, a supposed school, but it was actually a real-life death camp for them. Remarkable art teachers in the school defied camp regulations and offered art therapy, poetry, and literature to help the children deal with the horror of loss and death in their lives. The children, most of whom were going to die, were able to express their feelings and fears, and a humanness and life unequaled by the forces of brutality and death of their Nazi captors. This book shows the beauty of life, not the story of death, from a children's point of view. There is meaning here even if there is a sad ending.

It is the art that brings out the humanity of unknown children, all the way across time to us, and speaks loudly about any person or theory that denies our humanness. The children seemed to thrive in the midst of death and saw and gave us glimpses of what the mind could do. They were more than survivors in the evolutionary psychology scheme of things. They were persons and their brains were more than tools for survival. They were beings of hope, relationships, and symbolic existence, and whose words and thoughts could move them in time out of time from their barracks to the sun. Here is part of a poem written in 1944 by the children in Barracks L318 and L417, ages 10–16 years.

> The sun has made a veil of gold
> So lovely that my body aches.
> Above, the heavens shriek with blue
> Convinced I've smiled by some mistake.
> The world's abloom and seems to smile.
> I want to fly but where, how high?
> If in barbed wire, things can bloom
> Why couldn't I? I will not die![3]

And "The Butterfly" from Pavel Friedmann, written in 1942.

> For seven weeks I've lived in here,
> Penned up inside this ghetto.
> But I have found what I love here.
> The dandelions call to me
> And the white chestnut branches in the court.
> Only I never saw another butterfly.[4]

Who Are We?

Now that we are at this end point, let me ask, "Who are we, given our knowledge of the brain?" Who is Man, now that we have studied the brain? Let me attempt to answer the Psalmist's question, "What is man that thou are mindful of him?"

What words should I use to describe the mind/body stuff up there in our heads? Maybe no words will ever adequately describe that three-pound brain up there. That is no reason not to use words, but when you are dealing with what everyone calls the most complicated physical structure in the universe, I cannot be so arrogant as to think I got it with my few words.

Whether fMRIs or paintings, or Connectome highways or whole persons, or bottom-up or top-down thinking; all of it seems needed. Shame on you if you think you have the truth just because you blind guides have grabbed ahold of the elephant's tail. The whole person may be your best guide to understanding the brain, the box top on the 10,000-piece puzzle. Who would think of throwing the box top away and attacking the pieces only? Keep working on those empirical pieces in subjection to the whole truth, a truth painted in gold for humankind.

How is it that we study the brain maps and we so arrogantly think we have the human person described by our empiricism? Is it not obvious that you and I are more than PET scans correlating with movements and vocalizations and thoughts? I am not arguing for more study in the arts, but for a more serious look at human consciousness and self-consciousness. These we see deeply in our own experiences, in the world of deep thought and feeling in art and science and accomplishment and genius and religion.

Concluding Thoughts

We are the key. We carry it around with us, brain and personhood, matter and mind, glued together until death, for a short moment in eternity. Then comes the resurrection of the brain for the future of this wonderful personhood and an eternity to be persons in love relationships with our God and neighbors.

The apologetic from desire is still there no matter what neuroscience might say, the longing for the ultimate that we see in all these great persons. Where else but the brain do we find nostalgia, faith, agape, aesthetic appreciation, humor, irony, and awe? Remind us of how massively beautiful the brain is, and how interesting man is in his accomplishments and genius—that should help us interpret what we see in the brain. Not just action potentials and brain maps. The person is a whole, let the details lead you in that direction.

Hamlet's thoughts as he held Yorick's skull were of the special memories of the court jester who lived and laughed with him. It is interesting that most of our knowledge about early man comes from skulls—yes we have the arrow heads and pottery and clovis points around the fire pits—but how quickly we would change our minds if we could see the works of that man if he was given an education and opportunity. If aliens happened upon the earth many years ago and found Native Americans in tepees, with bows and arrows, would they be able to guess the future of that man 400 years later? Would they see the F-35 Hawke jet fighters and the International space station? It's all the same head and brain, same volume, same cells, same weight, same transmitters. It would not make sense for an alien years from now to come to our long deserted planet and dig up your skull and only study its volume and guess at your primitiveness without looking around at the many deserted buildings and books. How could they know us just by examining the skull? Only Hamlet can tell us truly about Yorick.

Yes the buildings and books would be giving top down direction to the study of the skull, but not imposed by a doctrine defending itself against all data, but by the obvious world around us.

****

> **"What a piece of work is a man, how noble in reason, how infinite in faculties, in form and moving how express and admirable, in action how like an angel, in apprehension how like a God! The beauty of the world, the paragon of animals...."**
>
> **—*Hamlet*, Act II, Scene 2**

SOME BOOKS I THINK YOU MIGHT LIKE

***Absence of Mind: The Dispelling of Inwardness from the Modern Myth of the Self*, by Marilynne Robinson.** This brain book by one of the leading authors of our time is well worth the effort to read because she delves expertly into the misleading tendencies of neuroscience when it comes to explaining the person.

***Me, Myself, and Why: Searching for the Science of Self*, by Jennifer Ouellette.** This well-written, entertaining book covers a broad spectrum of topics, from brain science to genetics, as the author searches for an understanding of the person and herself.

Endnotes

Chapter 1 The Human Brain: An Introduction to a Mystery

1. The figure 86 billion neurons, instead of the usual 100 billion so often stated, comes from Dr. Suzana Herculano-Houzel, a biologist at Vanderbilt University, who made an actual count of the nuclei of neurons in a brain soup of cortical neurons. She summarizes her work in, "The Remarkable (But Not Extraordinary) Human Brain," *Scientific American Mind*, March/April, 2017, pp. 36–41.

Chapter 2 Brain on the Table: The Anatomy of Mind

1. Mirror neurons were discovered by Dr. Rizzolatti, an Italian neuroscientist at the University of Parma in Italy. It is possible that these neurons play a role when we imitate the actions or emotions of someone and thus begin to understand the intent of those actions and feelings. Rizzolatti G, Craighero, L. "The mirror neuron system." *Annual Review of Neuroscience. 27* (1), 2004: 169–192.
2. C. O. Sylvester, ed. *Roget's Thesaurus of English Words and Phrases*, New York, Thomas Y. Crowell Col, 1911, p. x.

Chapter 3 Rivers of the Mind: Shaping the Self

1. See Gyorgy Buzsaki, *Rhythms of the Brain*. (New York: Oxford University Press, 2006), for a fuller understanding of these rhythms in the brain.
2. W. Denk. and K. Svoboda. "Photon upmanship: why multiphoton imaging is more than a gimmick." *Neuron* 18 (1997): 351–357.
3. Semir Zeki, *et.al.*, "The experience of mathematical beauty and its neural correlates." *Frontiers in Human Neuroscience. 13* Feb (2014). https://doi.org/10.3389/fnhum.2014.00068.
4. Steven Brown and Lawrence Parsons. "The neuroscience of dance." *Scientific American*, July (2008), 78–83.

Chapter 4 The Hard Problem: Neural Pixie Dust or God's Spirit

1. David J. Chalmers. *The Conscious Mind: In Search of a Fundamental Theory*. (Oxford University Press, 1996).
2. Thomas Nagel, "What Is It Like To Be a Bat?" *Philosophical Review 83* (1974): 435–450.
3. Ibid. 436.
4. Frank Lynn Meshberger, MD, "An Interpretation of Michelangelo's Creation of Adam, Based on Neuroanatomy." *JAMA Journal of the American Medical Association*, Oct 10, 1990, *Vol 264*, No. 14, p 1837–1841.
5. R. Quian Quiroga, *et al.* "Invarient visual representation by single neurons in the human brain." *Nature Vol 435*, 23 (June 2005): 1102–1107. Doi:10.1038/ nature03687.

Chapter 5 Free Will or Free Won't: Somewhat Free and Somewhat Not

1. Francis Crick, *The Astonishing Hypothesis: The Scientific Search for the Soul* (New York: Simon and Schuster, 1994), p. 3.
2. Eddy Nahmias. "Why we have free will," *Scientific American*, Jan (2015): 77–79.

3. Ibid., 77–79.
4. Marcel Brass and Patrick Haggard. "To Do or Not to Do: The Neural Signature of Self-Control. *Journal of Neuroscience, 34*, (22 August, 2007): 9141-9145; DOI: https: //doi. org/10.1523; JNEUROSCI. 0924-07. 2007.
5. H.C. Barron, *et. al.* "Unmasking Latent Inhibitory Connections in Human Cortex to Reveal Dormant Cortical Memories," *Neuron*, Vol 90, issue 1 (6 April, 2016): 191–203.
6. A good summary of John Polkinghorne's ideas can be found in *Belief in God in an Age of Science* (New Haven CT: Yale University Press, 1998).
7. Ibid., 14.
8. H. E. Fisher, *et. al.*, "Reward, addiction, and emotion regulation systems associated with rejection in love" *Journal of Neurophysiology*, July *104* (2010): 51–60.

Chapter 6 God Spots on the Brain: Putting God Back Where He Belongs

1. Alexander G. Huth, *et. al.*, "Natural speech reveals the semantic maps that tile human cerebral cortex." Nature, 532, (*28* April 2016), 453–458. doi:10.038/nature17637.
2. An often quoted sentence from Arthur C. Clarke. *Profiles of the Future: An Inquiry into the Limits of the Possible* (New York: Popular Library, 1973) 14.
3. Arianna Palmieri, et. al. "Reality of near-death-experience memories: evidence from a psychodynamic and electrophysiological integrated study." *Frontiers of Human Neuroscience, 8*: (2014) 429.
4. Mario Beauregard, and Vincent Paquette, "Neural correlates of a mystical experience in Carmelite nuns," *Neuroscience letters*, Sept 25, *405*, 3 (2006), 188–190.
5. Andrew Newberg, et. al. *Why God Won't Go Away: Brain Science and the Biology of Belief* (New York: Ballantine Books, 2002), 36.
6. Robert Coles. *The Spiritual Life of Children* (Boston: Houghton Mifflin Co: A Peter Davison Book, 1990).
7. Thomas Cahill, *How the Irish Saved Civilization* (New York: Anchor Books, Double Day Dell, 1995), 124.
8. Malcolm Jeeves and Warren S. Brown. *Neuroscience, Psychology, and Religion: Illusions, Delusions, and Realities about Human Nature* (West Conshohocken, PA: Templeton Press, 2009).

Chapter 7 Persons at the Edges of Personality: Still There, Just Hidden from View

1. Oliver Sacks, *A Leg to Stand On* (New York: Harper and Row, 1984), 124.
2. Oliver Sacks, *Awakenings* (New York: E. P. Dutton, 1983).

Chapter 8 Recreating the Human Being: Future Neuro-Technologies and Robots

1. Angelika Dimoka, "How to conduct a functional magnetic resonance (FMRI) study in social science research" *MIS Quarterly, 36*, no 3 (2012): 811–840.
2. Jose Delgado, *Physical Control of the Mind: Toward a Psychocivilized Society* (New York: Harper Colophon, 1971).
3. Ray Durzweil, "How Infinite in Faculty," *Discover Magazine,* November (2012), 54–55. Michio Kaku, *The Future of the Mind* (New York: Penguin Random House, 2015).

Chapter 9 The End of the Matter: Bottled Brains

1. J. W. Lichtman, and J. R. Sanes, "Ome sweet ome; What can the genome tell us about the connectome?" *Current Opinion Neurobiology, 18,* 3, (2008), 346–353.
2. Susan Blackmore, *Conversations On Consciousness* (New York: Oxford University Press, 2006) 49.
3. Ibid. 61.
4. Ibid. 61.
5. Ibid. 77.
6. Ibid. 196.

Chapter 10 Thoughts about Persons and Brains: You Were Never out of My Mind

1. R. W. Franklin, ed., *The Poems of Emily Dickinson*; (Cambridge: Belknap Press, 1999).
2. T. S. Eliot, from the opening stanza of "The Rock" (London: Faber and Faber, 1934).
3. Hana Volavkova, Ed., "*I Never Saw Another Butterfly: Children's Drawings and Poems from Terezin Concentration Camp 1942-1944.*" (New York: Schocken Books, 1993) 39.
4. Ibid. 77.

For Further Reading on the Subjects in this Book

Ackerman, D. (2005). *An Alchemy of Mind: The Marvel and Mystery of the Brain*. Scribner.

Baker, M. C., & Goetz, S. (Eds.). (2010). *The Soul Hypothesis: Investigations into the Existence of the Soul*. Continuum.

Beauregard, M. (2012). *Brain Wars: The Scientific Battle over the Existence of the Mind and the Proof that will Change the Way We Live Our Lives*. New York: HarperOne.

Beauregard, M., & O'Leary, D. (2008). *The Spiritual Brain: A Neuroscientist's Case for the Existence of the Soul* (Reprint edition). HarperOne.

Blackmore, S. (2005). *Consciousness: A Very Short Introduction* (1st edition). Oxford, UK ; New York, N.Y: Oxford University Press.

Blackmore, S. (2006). *Conversations on Consciousness: What the Best Minds Think about the Brain, Free Will, and What it Means to be Human*. Oxford ; New York: Oxford University Press.

Carter, R. (2010). *Mapping the Mind* (Revised edition). Berkeley: University of California Press.

Chalmers, D. J. (1996). *The Conscious Mind: In Search of a Fundamental Theory*. Oxford University Press.

Churchland, P. S. (2014). *Touching a Nerve: Our Brains, Our Selves*. W. W. Norton & Company.

Crick, F. (1994). *The Astonishing Hypothesis: The Scientific Search for the Soul*. New York: Charles Scribner's Sons.

Custance, A. C. (1979). *The Mysterious Matter of Mind*. Grand Rapids, Mich: Zondervan.

Damasio, A. (2005). *Descartes' Error: Emotion, Reason, and the Human Brain*. Penguin Books.

Damasio, A. (2012). *Self Comes to Mind: Constructing the Conscious Brain*. New York: Vintage.

Dehaene, S. (2014). *Consciousness and the Brain: Deciphering How the Brain Codes Our Thoughts*.

Delgado, J. M. R. (1971). *Physical Control of the Mind—Toward a Psychocivilized Society* (Second edition). New York: Harper & Row.

Dietrich, A. (2007). *Introduction to Consciousness*. Basingstoke; New York: Palgrave.

Doidge, N. (2007). *The Brain that Changes Itself: Stories of Personal Triumph from the Frontiers of Brain Science*. New York: Penguin Books.

Duhigg, C. (2014). *The Power of Habit: Why We Do What We Do in Life and Business*. New York: Random House Trade Paperbacks.

Eagleman, D. (2012). *Incognito: The Secret Lives of the Brain*. New York: Vintage.
Eagleman, D. (2015). *The Brain: The Story of You*. Pantheon.
Eccles, J. (1985). *The Wonder of Being Human*. Boston : New York: Shambhala.
Eccles, J. C. (1994). *How the Self Controls Its Brain*. Berlin, Heidelberg: Springer.
Eccles, J. C., & Popper, K. (2014). *The Self and Its Brain: An Argument for Interactionism* (Revised ed. edition). Routledge.
Edelman, G. M. (2007). *Second Nature: Brain Science and Human Knowledge*. Amsterdam ; New York: Yale University Press.
Edelman, G. M., & Tononi, G. (2000). *A Universe of Consciousness: How Matter Becomes Imagination*. New York, NY: Basic Books.
Eliot, T. S. (2014). *Four Quartets*. Mariner Books.
Evans, C. S. (1994). *Preserving the Person*. Regent College Publishing.
Gazzaniga, M. S. (2009). *The Ethical Brain*. Dana Press.
Gordon, D. (Ed.). (2010). *Cerebrum 2010: Emerging Ideas in Brain Science* (1st Edition edition). New York: Dana Press.
Grandin, T. (2006). *Thinking in Pictures, Expanded Edition: My Life with Autism*. New York: Vintage.
Graziano, M. S. A. (2010). *God Soul Mind Brain: A Neuroscientist's Reflections on the Spirit World*. Teaticket, Mass: Leapfrog Press.
Green, J. B. (Ed.). (2004). *What about the Soul?* Nashville: Abingdon Press.
Hooper, J., & Teresi, D. (1986). *The Three-Pound Universe*. New York: Macmillan.
Horstman, J., & American, S. (2010). *The Scientific American Brave New Brain: How Neuroscience, Brain-Machine Interfaces, Neuroimaging, Psychopharmacology, Epigenetics, the Internet, and ... and Enhancing the Future of Mental Power*. San Francisco, Calif: Jossey-Bass.
Howard, T. (2006). *Dove Descending: A Journey into T.S. Eliot's Four Quartets*. Ignatius Press.
Humphrey, N. (2011). *Soul Dust: The Magic of Consciousness*. Princeton University Press.
Jeeves, M. (Ed.). (2004). *From Cells to Souls—and Beyond: Changing Portraits of Human Nature*. Grand Rapids, Mich: Wm. B. Eerdmans Publishing Company.
Jeeves, M., & Brown, W. S. (2009). *Neuroscience, Psychology, and Religion: Illusions, Delusions, and Realities about Human Nature*. West Conshohocken, Pa: Templeton Press.
Kagan, J. (2007). *An Argument for Mind* (1 edition). Yale University Press.
Kahneman, D. (2013). *Thinking, Fast and Slow*. New York: Farrar, Straus and Giroux.
Kaku, M. (2014). *The Future of the Mind: The Scientific Quest to Understand, Enhance, and Empower the Mind*. Doubleday.
Kellogg, R. T. (2013). *The Making of the Mind: The Neuroscience of Human Nature*. Prometheus Books.

Koch, C. (2012). *Consciousness: Confessions of a Romantic Reductionist.* Hardcover. (n.d.). MIT Press.
Koch, C. (2004). *The Quest for Consciousness: A Neurobiological Approach.* Denver, Colo.: W. H. Freeman.
Koob, A. (2009). *The Root of Thought: Unlocking Glia—The Brain Cell that Will Help Us Sharpen Our Wits, Heal Injury, and Treat Brain Disease.* Pearson FT Press.
Kosslyn, S. M. (2010). *Wet Mind: The New Cognitive Neuroscience.* Free Press.
Kurzweil, R. (2013). *How to Create a Mind: The Secret of Human Thought Revealed.* New York: Penguin Books.
Loewenstein, W. (2013). *Physics in Mind: A Quantum View of the Brain.* New York: Basic Books.
Marcus, G., & Freeman, J. (Eds.). (2014). *The Future of the Brain: Essays by the World's Leading Neuroscientists.* Princeton University Press.
Matthews, P. M., & McQuain, J. (2003). *The Bard on the Brain: Understanding the Mind Through the Art of Shakespeare and the Science of Brain Imaging.* New York: Dana Press.
MD, J. M. S., & Begley, S. (2003). *The Mind and the Brain: Neuroplasticity and the Power of Mental Force.* Harper Perennial.
Medina, J. (2014). *Brain Rules:: 12 Principles for Surviving and Thriving at Work, Home, and School.* Pear Press.
Museum, U. S. H. M. (1994). *I Never Saw Another Butterfly: Children's Drawings and Poems from the Terezin Concentration Camp, 1942-1944.* (H. Volavkova, Ed.) (2nd edition). Schocken.
Nagel, T. (2012). *Mind & Cosmos: Why The Materialist Neo-Darwinian Conception of Nature is Almost Certainly False.* New York: Oxford University Press.
Newberg, A., D'Aquili, Eugene, Rause, Vince. (n.d.). *Why God Won't Go Away: Brain Science and the Biology of Belief.* Ballantine Books, 2002.
Nuland, S. B. (1995). *How We Die: Reflections of Life's Final Chapter.* New York: Vintage.
O'Shea, M. (2005). *The Brain: A Very Short Introduction.* OUP Oxford.
Ouellette, J. (2014). *Me, Myself, and Why: Searching for the Science of Self.* Penguin Books.
Penrose, R. (1996). *Shadows of the Mind: A Search for the Missing Science of Consciousness.* New York: Oxford University Press.
Polkinghorne, J. (1998). *Belief in God in an Age of Science* (Revised ed. edition). Yale University Press.
Pribram, K. H. (2013). *The Form Within: My Point of View.* Prospecta Press.
Ramachandran, V. S. (2011). *The Tell-Tale Brain: A Neuroscientist's Quest for What Makes Us Human.* W. W. Norton & Company.
Restak, R. M. (2016). *The Big Questions: Mind.* London: Quercus Publishing.
Richtel, M. (2014). *A Deadly Wandering: A Tale of Tragedy and Redemption in the Age of Attention.* New York, NY: William Morrow.

Robinson, M., & Dwight Harrington Terry Foundation. (2010). *Absence of Mind: The Dispelling of Inwardness from the Modern Myth of the Self.* New Haven: Yale University Press.

Sacks, O. (1974). *Awakenings,*. Garden City, NY: Doubleday.

Sacks, O. (1996). *An Anthropologist on Mars: Seven Paradoxical Tales*. New York: Vintage Books.

Sacks, O. (1998). *A Leg to Stand On* (Reprint edition). Touchstone.

Sacks, O. (2008). *Musicophilia: Tales of Music and the Brain, Revised and Expanded Edition*. New York: Vintage.

Sacks, O. (2010). *The Man Who Mistook His Wife for a Hat and Other Clinical Tales*. Odyssey Editions.

Sacks, O. (2015). *Gratitude*. Knopf.

Sagan, C. (1979a). *Broca's Brain*. New York: Random.

Sagan, C. (1979b). *The Dragons of Eden* (2nd Impression edition). Coronet Books.

Seung, S. (2012). *Connectome: How the Brain's Wiring Makes Us Who We Are*. Boston: Houghton Mifflin Harcourt.